Studies of the Americas

edited by
Maxine Molyneux
Institute for the Study of the Americas
University of London
School of Advanced Study

Titles in this series are multidisciplinary studies of aspects of the societies of the hemisphere, particularly in the areas of politics, economics, history, anthropology, sociology, and the environment. The series covers a comparative perspective across the Americas, including Canada and the Caribbean as well as the United States and Latin America.

Titles in this series published by Palgrave Macmillan:

Cuba's Military 1990–2005: Revolutionary Soldiers during Counter-Revolutionary Times
By Hal Klepak

The Judicialization of Politics in Latin America
Edited by Rachel Sieder, Line Schjolden, and Alan Angell

Latin America: A New Interpretation
By Laurence Whitehead

Appropriation as Practice: Art and Identity in Argentina
By Arnd Schneider

America and Enlightenment Constitutionalism
Edited by Gary L. McDowell and Johnathan O'Neill

Vargas and Brazil: New Perspectives
Edited by Jens R. Hentschke

When Was Latin America Modern?
Edited by Nicola Miller and Stephen Hart

Debating Cuban Exceptionalism
Edited by Bert Hoffman and Laurence Whitehead

Caribbean Land and Development Revisited
Edited by Jean Besson and Janet Momsen

Cultures of the Lusophone Black Atlantic
Edited by Nancy Priscilla Naro, Roger Sansi-Roca, and David H. Treece

Democratization, Development, and Legality: Chile, 1831–1973
By Julio Faundez

The Hispanic World and American Intellectual Life, 1820–1880
By Iván Jaksić

The Role of Mexico's Plural *in Latin American Literary and Political Culture: From Tlatelolco to the "Philanthropic Ogre"*
By John King

Faith and Impiety in Revolutionary Mexico
Edited by Matthew Butler

Reinventing Modernity in Latin America: Intellectuals Imagine the Future, 1900–1930
By Nicola Miller

The Republican Party and Immigration Politics: From Proposition 187 to George W. Bush
By Andrew Wroe

The Political Economy of Hemispheric Integration: Responding to Globalization in the Americas
Edited by Diego Sánchez-Ancochea and Kenneth C. Shadlen

Ronald Reagan and the 1980s: Perceptions, Policies, Legacies
Edited by Cheryl Hudson and Gareth Davies

Wellbeing and Development in Peru: Local and Universal Views Confronted
Edited by James Copestake

The Federal Nation: Perspectives on American Federalism
Edited by Iwan W. Morgan and Philip J. Davies

Base Colonies in the Western Hemisphere, 1940–1967
By Steven High

Beyond Neoliberalism in Latin America? Societies and Politics at the Crossroads
Edited by John Burdick, Philip Oxhorn, and Kenneth M. Roberts

Visual Synergies in Fiction and Documentary Film from Latin America
Edited by Miriam Haddu and Joanna Page

Cuban Medical Internationalism: Origins, Evolution, and Goals
By John M. Kirk and H. Michael Erisman

Governance after Neoliberalism in Latin America
Edited by Jean Grugel and Pía Riggirozzi

Modern Poetics and Hemispheric American Cultural Studies
By Justin Read

Youth Violence in Latin America: Gangs and Juvenile Justice in Perspective
Edited by Gareth A. Jones and Dennis Rodgers

The Origins of Mercosur
By Gian Luca Gardini

Belize's Independence & Decolonization in Latin America: Guatemala, Britain, and the UN
By Assad Shoman

Post-Colonial Trinidad: An Ethnographic Journal
By Colin Clarke and Gillian Clarke

Post-Colonial Trinidad

An Ethnographic Journal

Colin Clarke and Gillian Clarke

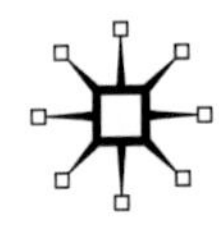

First published in 2010 by
PALGRAVE MACMILLAN®
in the United States—a division of St. Martin's Press LLC,
175 Fifth Avenue, New York, NY 10010.

Where this book is distributed in the UK, Europe and the rest of the world, this is by Palgrave Macmillan, a division of Macmillan Publishers Limited, registered in England, company number 785998, of Houndmills, Basingstoke, Hampshire RG21 6XS.

Palgrave Macmillan is the global academic imprint of the above companies and has companies and representatives throughout the world.

Palgrave® and Macmillan® are registered trademarks in the United States, the United Kingdom, Europe and other countries.

ISBN: 978–0–230–62200–5

Library of Congress Cataloging-in-Publication Data is available from the Library of Congress.

A catalogue record of the book is available from the British Library.

Design by Newgen Imaging Systems (P) Ltd., Chennai, India.

First edition: May 2010

10 9 8 7 6 5 4 3 2 1

Printed in the United States of America.

To the memory of
Bramadath Maharaj, Hansar Ramsamooj, George Sammy,
and Ena Scott-Jack,
and with gratitude to all who assisted
with the fieldwork

Contents

Figures

Plates

Glossary

The nonstandard English terms in the glossary have been checked against Lise Winer's *Dictionary of the English/Creole of Trinidad and Tobago*, 2009. However, as there is no published dictionary of *Bhojpuri* (in Trinidad originally thought to have been Hindi, as our informants indicate), I have selected one of the several orthographies given by Winer for words in that language.

aarti Hindu worship, a ritual in which a small fire in a flat container or lamp is moved in circular motion in front of a god or person

abir purple dye thrown on celebrants during Holi or Phagwa

aguwah marriage broker

agwani meeting up of bride's and groom's fathers at the beginning of an Arya Samaji wedding, during which the fathers garland one another

aji father's mother

ajoupa house with wattle-and-daub walls and thatched roof of dried palm leaves

akika Muslim sacrificial rite held seven days after a child's birth

aloo potato

anchar a sweet or salty peppery preserve made from green unripe fruit, usually mangoes

ashram abode or hall attached to Hindu temple or *mandir*

badnazur evil eye

Bakra Eid festival celebrating the story of Abraham's forestalled sacrifice of Isaac (also known as Eid ul Azha)

bap father

barahi 12 days after a Hindu child's birth

barat motorcade, involving Hindu groom and his entourage as they proceed to the home of the bride; traditionally they would have ridden on a horse

barka-chotki husband's eldest brother/wife, an avoidance relationship

barka mai big aunty

Bhagwat Yagya weeklong ceremony, involving the reading and exposition of Bhagvata-Purana or the Bhagavad-Gita

bhajan Hindu hymn or holy song with words in Hindi

bhandari to make *persad*

birha traditional Indian songs sung at Hindu wedding

black African (of slave origin in the Caribbean)

bobol corruption, graft, or fraud

brown mixed racial group of white-black origin

channa a cultivated edible pea or chick pea

charidar gather flowers and make garlands

chati sixth day after birth of Hindu child

chauk altar

chela initiate or pupil of a spiritual leader

Christian East Indian Indian Christian convert from Islam or Hinduism, or a descendant

churkee topknot on shaven head

class a socioeconomic grouping (class) ranked in power, prestige, and consumption capacity from low to medium to high class

color-class a common feature of Caribbean societies is a close relationship between color and class, so that white, brown, and black coincide closely with upper-, middle- and lower-class categories

color(ed) mixed race group, between black and white

coolie term applied to Asian indentured laborers on the Caribbean sugar plantations, here East Indian

coolie English English Creole spoken by East Indians

crapaud French Creole for toad

Creole a person born in the Caribbean, but of foreign descent; a term that distinguishes white, colored, and black West Indians (the categories created and ranked during plantation slavery) from all others, notably East Indians

Creole Trinidad language variant of English and French, hence English Creole and French Creole

culture behavior associated with distinct religious, familial, or educational beliefs and practices, or a combination of them

dada father's elder brother, or equivalent

dadi father's elder brother's wife, or equivalent

dal lentil, split pea

dal puja garlanding ceremony during Hindu wedding, as the groom's elder brother makes gifts to the bride

dalpuri a flat Indian bread, like a *roti*, with a *dal* (lentil) filling

deota Hindu god

dewan secretary

dhantal iron bar used for percussion

dharma religious duty

dholak cylindrical Indian drum, played at both ends with hands

dhoti loincloth with appearance of loose trousers

dia small, red clay lamp, filled with oil and lit with a cotton wick

didi older female sibling or cousin

Diwali festival of lights, devoted to Lakshmi, and celebrating the return from the forest of Lord Rama and his wife Sita

doolaha bridegroom

doolahin bride

dougla (Hindi for hybrid) meaning an East Indian and black mixture (pronounced doo-gla)

Durga puja ceremony dedicated to goddess Durga to ward off sickness

Dwar puja ceremony of the gateway at Hindu weddings

East Indian indentured laborer from India and descendants

Eid ul Fitr feast ending fast of *Ramadan*, also known as *Eid ul Doh*. First day of Muslim New Year

ethnicity group defined by culture (such as religion or language) and history

Ganesh Hindu god with elephant head

ganja *cannabis sativa* (marijuana)

ghangri traditional long skirt worn by indentured Indian women in the Punjab and western parts of Uttar Pradesh

ghee clarified butter

gobar cow dung, or a mixture of cow dung and mud used to plaster floors and walls

Gobardhan puja prayer in honor of Gobardhan Mountain, which takes place on the Thursday evening of a Rameyn Yagya or a Bhagwat Yagya

gotra clan, based on paternal inheritance

gupta dan "gift in secret"—ritual during Hindu wedding when a ball of dough, containing money, is given to the groom by the bride and her parents

guru Hindu spiritual teacher or guide

gurudiksha Hindu ceremony when a person is given a personal guide, usually a pundit

gurumukh initiation ceremony or "christening" when a special *mantra* is given to a Hindu man or woman

hafiz Muslim able to recite the Koran by heart; a Muslim who officiates at a wedding ceremony

Haj pilgrimage to Mecca

haji someone who has made the pilgrimage to Mecca.

halwa sweetmeat of flour, ghee, and sugar

Hanuman Hindu monkey god, also known as Mahabir, whose puja is *rōt*

haris a piece of mango wood, dyed yellow; it usually has five or seven notches cut into it

hawan Hindu ceremony of feeding a sacred fire, usually involving ritual oblations of *ghee*

Hindu Indian believer in various gods ranging from the village to the Sanskritic deities

Holi or Phagwa Hindu spring festival, celebrating the new year; day of secular fun in late March, known in Trinidad as "Indian carnival"

Hosay Shiite festival to commemorate martyrdom of Hosain, Mohammed's grandson, AD 680

imam Muslim priest

jab molassi Carnival devil covered with molasses

jahaaj ship

jahaji bhai shipmate friend, usually male

jajmani system of traditional economic exchange between castes

jal holy water, poured as offering during *puja*

jamaat congregation of Muslims

janeo Hindu sacred thread, taken by the three (upper) twice-born varnas—Brahmin, Kshatriya, and Vaishya

Jhanam Astamie birth of Lord Krishna

jhandi triangular flag on long bamboo pole associated with Hindu ritual.

jharay "sweeping away" of sickness, involving stroking and prayers

*jora-jama (*also *jama-jora)* Indian man's long robe, usually pink, worn by the groom at a Hindu wedding

j'ouvert French Creole for *jour ouvert, or* morning of Carnival Monday (preceding Ash Wednesday)

jhula loosely fitting bodice, with arms up to wrist

jumbie spirit of a dead person or ghost

jump-up dance—move vigorously in time to music

kajar soot from the burning of ghee, rubbed on the eyelids of new-born Muslim babies to beautify the eyes or keep away *maljeu*

kala pani black water (sea)

kalima faith (Muslim)—there is no god but Allah and Mohammed is his prophet

Kali Mai ceremony devoted to Kali, the Hindu goddess of death and destruction

kanya dan "virgin giving," part of the Hindu wedding ceremony when the father gives the bride to the groom

kardhan string tied around the stomach of a baby or child, usually male

katha reading from sacred Hindu text, includes prayers and singing

kichree food served to groom and his kinsmen at Hindu wedding; eating this signifies satisfaction with the gifts they have received

kitab ceremony of readings from the Koran with songs and a sermon

kohabar retiring room used by bride and groom once Hindu wedding is over

Koran holy book of Islam, revealed by God to Mohammed

kowsil to observe mistakes and make judgments

Krishna incarnation of Vishnu

kujat outcaste

kurta loose-fitting shirt or tunic with long sleeves, usually collarless, worn by Hindu men

kutia small Hindu shrine or temple

La Divina Pastora see Siparu Mai

Lakshmi Goddess of Light and Knowledge

lawa parched rice, used in Hindu wedding rituals

likanimul writer of books

lingam phallic stone devoted to Shiva

lotah round brass vessel for containing water

Mahadeo stone *lingam* or phallic stone

mahant low-caste priest

maktab Islamic school, teaching Urdu and Arabic

mala garland of fresh flowers

maljeu evil eye

mandir Hindu temple

manjira brass cymbals

mantra sacred Hindu text, used as prayer or ritual utterance

Mardi Gras Shrove Tuesday

maro Hindu marriage booth

maro hilai shaking of the marriage pole by bridegroom's father as mark of his satisfaction

mas' Carnival costume or character (masquerade)

matti kore planting of marriage pole, Hindu wedding, three days before event; mud or clay is specially dug by women while performing sexually explicit songs *maur* crown worn by groom at Hindu wedding

milap meeting-up of the fathers of the bride and groom and their respective entourages prior to Hindu wedding

Mouloud Sharif Muslim celebration involving readings from the Koran, religious songs, and a sermon

muezzin person who calls the Muslim faithful to prayer

mulki/mulkin term used to refer to inhabitant of India, male and female

Muluk the term used to refer to India

murti image, usually a small statue of a Hindu god

Muslim believer in divinity of Allah and prophecy of Mohammed

namaaz Muslim prayers, offered five times a day

nama sanskar Hindu postbirth ritual nail cutting and ablutions for baby and mother

namaste spoken Hindu greeting, especially among Trinidad Arya Samajis

nana maternal grandfather

nani maternal grandmother

Nau barber caste. Nau's wife (often played by a woman of another caste) is important in marriage ritual at Hindu weddings

neuta washing of feet of bride and groom at Hindu wedding

obeah black magic, witchcraft

ojha man Hindu practitioner of Indian folk rituals for divination and black magic

Om word without ending—eternity

oronhi traditional Indian woman's scarf or veil worn by married Muslim, Christian, and Hindu women

pan betel leaf

panchayat meeting of older men, village, or caste council

pani graham the groom's acceptance of the bride's hand in Arya Samaj weddings

parata roti special roti for Muslims, usually torn into pieces before eating

parchan ritual greeting of the groom at the bride's home, Hindu wedding, following *dwar puja*

parishad council, usually of pundits

patra astrological horoscope used by Hindu priests to select favorable days for events

pau puja gifting ceremony during Hindu wedding, when bride's father and relatives wash groom's feet

pelau chicken, rice, and pigeon peas

persad sacred Hindu food, made of *ghee*, sugar, flour, raisins, and coconut

Phagwa (aka Holi) Hindu spring festival, celebrating the new year; day of secular fun in late March, known in Trinidad as 'Indian carnival'

phuphu father's sister or equivalent kin

picong teasing, ridicule, or insult

pirah Hindu wedding bench made from a single piece of wood

puja prayer ceremony devoted to Hindu deity

pujari celebrant of *puja*

pundit Hindu priest

race social category defined with reference to the physical characteristics of its members in comparison with others in society

Radha consort of Krishna

rakaat genuflections performed during Muslim prayer cycle

Rama Lord and hero of the Ramayana, where he is treated as an incarnation of Vishnu

Ramadan monthlong fast during daylight hours enjoined on orthodox Muslims

Ramayana Hindu epic of Lord Rama, Sita and Hamunan, told by Tulsidas

Rameyn Satsang sung performance of the Ramayana, with readings and prayers

Ramlila Rama pageant

Ram Naumi celebration of Rama's birth

rehal hinged book rest, carved from a single piece of wood

Rishi Bodh Utsov Arya Samaji term for Siw Ratri

rōt Hanuman puja

roti round, flat, unleavened bread

roza fasting (Muslim)

saddhu Hindu holy man, usually older, who abstains from meat and alcohol *sahar* town or more specifically Port of Spain

Sanathan Dharma Maha Sabha Hindu religious organization: the most orthodox of Hindu sects under Brahmin control

sandhya puja daily prayers of the "twice-born" Hindus; meditations to deities in evening prayers with *hawan*

sanskar 16 ritual sacraments of Hinduism, focusing on rites of passage such as birth and death

saptapadi the seven steps around the sacred fire that seal the Hindu marriage

Satnaryn (Satyanarayan) Hindu god of truth

satsang Hindu religious gathering; includes prayers, chanting, and discourse

Shuddi Sanskar initial purification of Arya Samaji candidate

siballa companion (best man), Hindu wedding

sindoor vermillion powder rubbed into the central hair parting of Hindu woman to indicate her married status; also used to make *tika* mark on her forehead

sindoor dan rubbing of the *sindoor* into parting of the wife's hair, Hindu wedding ceremony

Siparu Mai Siparia Mother or La Divina Pastora, the black virgin of Siparia

siwala Hindu temple dedicated to Shiva

Shiva God the destroyer

Siw Ratri celebration of Shiva's birth

sloka stanza

souse marinated pig's trotters

surma black dust used by Muslims on eyes of children

Suruj Puran (Suruj narayan) Hindu sun god

swami Hindu religious leader

tabla drums played with the fingers, usually paired as bass and tenor

tadjah wood, paper, and tinsel model of the tomb of Hosain, constructed for Hosay (a Shiite celebration) to commemorate his martyrdom

tante French Creole for aunt

taraweeh prayers said at night during Ramadan

tarriah circular brass tray, used in Hindu rituals

tawa flat, circular metal griddle for cooking *roti*

tazim Muslim song of welcome

tika spot made on forehead as sign of religious devotion, betrothal, or ornament

tulsi tree cultivated bush of sacred basil plant

varna grouping of equivalent Hindu castes—Brahmin, Kshattrya, Vaishya, and Sudra

Veda sacred knowledge or book; one of the four sacred books of the Hindus

vedi Hindu altar made of earth, decorated in various colors for religious ceremonies

vedic relating to the Vedas; in Trinidad implies Arya Samaj

Vishnu god the preserver

white group of European origin

yagya/yagna literal meaning is sacrifice; in Trinidad term refers to a seven- or fourteen-day series of rites and ceremonial readings from a sacred text

yogi a practitioner of yoga

zakat compulsory charity, Muslim

Introduction

Colin Clarke

This ethnographic field journal was kept by Gillian and me during 1964, and its subject matter covers the second and third years of independence of the former British Caribbean colony of Trinidad and Tobago.[1] The main reason for our research was the widespread international concern that Trinidad's plural society,[2] comprising urban Creoles and rural East Indians (for definitions see glossary, pages xiii–xxii), would disintegrate through race/ethnic rivalry after it achieved sovereignty in 1962. More than forty-five years on, after a decade of greater cross-race collaboration, Trinidadians have recently reverted to the same race-competitive system that characterized the end of colonialism. The relevance of this book is, therefore, both historical and contemporary.

How had my involvement with research in Trinidad come about? In the summer of 1962, while working for my doctorate at Oxford University, I had received a letter from the Jamaican anthropologist, Professor M. G. Smith of the University of California, Los Angeles, mentor of my field work in Jamaica, asking me to submit a proposal focusing on urban East Indians in Trinidad to the Research Institute for the Study of Man (RISM) in New York. Dr. Vera Rubin, the director of RISM, reviewing recent research on Trinidad, had noted that rural studies had emphasized East Indian isolation, cultural retention, and hostility to Creole domination (Rubin, 1962). Were urban East Indians the same? This was *the* crucial question, given the pervasiveness of race voting in the electoral victories of 1956 and 1961[3] achieved by the black and brown Creole population.

Britain's preference for the colony of Trinidad and Tobago was not independence as a unitary state. Britain had anticipated that the two islands would form a unit of the Federation of the West Indies, due to become independent in 1962. It was the rejection of the federation by Jamaica in a referendum in 1962, followed by a general election in

Trinidad later in the year that led to Trinidad's withdrawal from the federation and its immediate quest for independence as a ministate. Though Trinidad was oil-rich and the only British Caribbean colony with substantial nonagricultural resources, with just 828,000 inhabitants, it had less than the magic figure of 1 million population that was set as the boundary below which state "viability" was thought to be unlikely.

Professor Smith's letter to me had suggested that I should involve Gillian, my wife, in the project (we had married in 1962), in the same way that he had engaged the assistance of his wife, Mary, in the field research for his doctorate in northern Nigeria in 1949–50 (Hall, 1997). So, in early January 1964, Gillian and I set out for a week's library research and briefings from Dr. Vera Rubin and Professor Lambros Comitas at RISM in New York, followed by short visits to Jamaica, Puerto Rico, and Antigua en route to Trinidad, which we reached shortly before the end of the month.

Prior to our departure from the United Kingdom, we had already purchased the 1960 census data for Trinidad and Tobago. These showed that San Fernando was the only urban settlement in which East Indians made up just over a quarter of the inhabitants and so there was virtually no choice other than to select it as our field site. However, with a population of almost 40,000—over half of whom were migrants—San Fernando appeared manageable for a full-time research team of two, and small enough for urban research that was to blend social geography, sociology, and social anthropology. San Fernando was the second town of Trinidad, after the capital, Port of Spain, and, with the sugar and oil industries surrounding it, was arguably the industrial center of the island.

I anticipated that the project would draw on three sources of information, only one of which (small-area census data for detailed mapping) was already available. The other two sources would have to be generated by us through our own fieldwork. The major dataset that we had to create was a questionnaire survey of five of San Fernando's principal racial and ethnic categories: (1) Creoles (white, brown, and black), (2) Hindus, (3) Muslims, and (4) Christian East Indians, and (5) mixed race (Indian/black) known as *douglas* (Hindi for hybrid; the singular is pronounced *doo-gla*), in addition to a sample of rural East Indians drawn from the adjacent village of Débé (see glossary for terms relating to race/color, religion, and ethnicity). Devising the questionnaire, and applying it to almost 900 respondents, using ten trained field assistants, occupied most of the second half of our research period in San Fernando.

The third source of information was to be the journal that we kept, more or less on a daily basis, during our stay in Trinidad. Through participant observation of a multiplicity of rituals and events, and a series of informal and semistructured interviews carried out over a nine-month period, we constructed a detailed picture of social life in San Fernando and the East Indian villages in the surrounding sugar belt and beyond by recording scenes and conversations relevant to our research aims. The journal was crucial for us in helping to identify the topics and issues that formed the basis for the questionnaire, and it became a repository of information and opinion that was drawn upon as background to my book, *East Indians in a West Indian Town: San Fernando, Trinidad, 1930–70* (Clarke, 1986).

However, only eight extracts in *East Indians in a West Indian Town* were taken directly from the journal, partly because, as a piece of social science, the book drew heavily on the cartographic and questionnaire data for San Fernando and Débé, and partly because the journal had a wider focus than the town and the village, but, above all, because the manuscript of the journal ran to over 100,000 words. It was impossible to incorporate so much material, other than by a brief citation, in an academic monograph, such as this, with a restricted word limit.

Journal entries were made to record everything that had social-scientific relevance to the project. The account was kept by both Gillian and me, though Gillian made most of the entries in the first half of the text, and I recorded the bulk of the interviews in the second half. No tape recordings were used. Gillian and I normally wrote the journal from memory in the early morning following an interview or an event; most of my interviews with individuals in the second half of the project were preserved in detailed notes written at the time of the meeting, with the permission of the respondent. Whenever possible, Gillian and I read each other's entries for completeness and accuracy of recall, and amended the document so that events and interviews, which we had shared, reflected our joint participation. This was more necessary and feasible in the first half of the project, when we were feeling our way, but had more time to devote to the journal.

Academic Context of the Field Research

Trinidad and Tobago is a two-island unitary state located in the Southeast Caribbean, off the east coast of Venezuela. Colonized by Spain from 1498 to 1797, Trinidad then became a British possession until its independence in 1962. On acquiring sovereignty, the Creole government set out to project a black image for the society—particularly by portraying Carnival as *the* national festival—to creolize the state, and to emphasize racial and religious harmony. "All o' we is one," runs a local saying, and the National Anthem reiterates, "Here every creed and race find an equal place." Nevertheless, differences of race, color, class, and culture (see glossary) have a long and complex history in Trinidad, and each of these defines important elements in the island's social structure.

Of overriding significance is the dichotomy between Creole and East Indian or, since independence, Indian. Both terms—East Indian and Indian—are used interchangeably in the journal, though among the young, there is now a growing preference for Indo-Trinidadian (Rampersad, 2002). The term "Creole" has a special meaning in Trinidad. It excludes the Indian population, together with the small Carib, Syrian, Portuguese, and Chinese minorities. Creoles may be white, brown, or black, but color still correlated, at the end of the colonial period, with their cultural and socioeconomic stratification.

Historically, stratification of the Creole segment was established during slavery, under the influence of white sugar planters and administrators (Wood, 1968). Slave emancipation in the period 1834–38 removed the legal distinctions between the strata, but left the old social hierarchy essentially intact. Within less than a decade, Indian indentured laborers (the men bonded for five years and the women for three) were brought in to work on the sugar estates, which, as in the case of British Guiana, were located on an expanding sugar frontier. During the seventy years of indentured immigration, 144,000 Indians came to Trinidad, of whom 110,000 never returned. Their descendants now account for just over two-fifths of the island population, though they are divided by religion into discrete subsegments.

Detailed social-scientific accounts of black communities in Trinidad date from the classic study of the village of Toco on the north coast (fig. 1), published by the Herskovits, U.S. anthropologists, in 1947 (Herskovits and Herskovits, 1964 edition). This was one of many sites they researched in the Caribbean in their quest for "Africanisms" among the descendants of the formerly enslaved black populations. The first detailed account of Trinidad's social structure by a professionally trained local sociologist, Lloyd Braithwaite, was published in 1953; however, it focused almost entirely on the Creole component of the society and explored its late colonial, color-class stratification (Braithwaite, 1953).

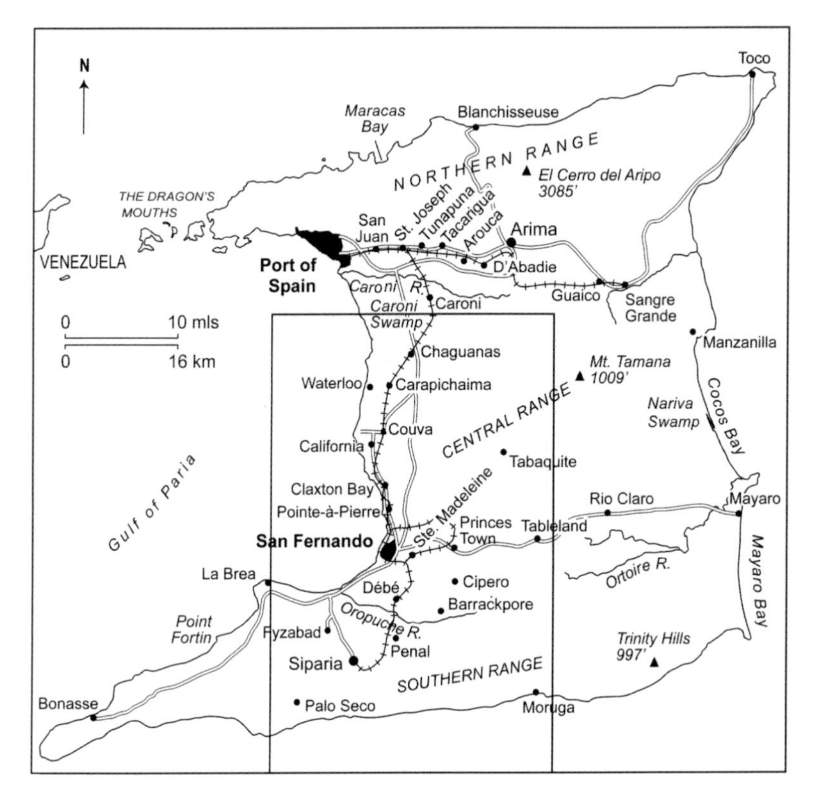

Figure 1 Trinidad: selected place-names and communications in the 1960s

Braithwaite confined his comments on the East Indians to the observation that "the rise of a vocal East Indian middle class and the elimination of the more demoralized type of free Indians, who gave the 'creole' public the stereotype of a 'coolie,' present a problem of major importance for the social structure" (Braithwaite, 1953, 9). In 1960, however, Braithwaite added that Creole "values are, however, not so firmly implanted among the Indian sector (35 percent of the population) largely because of the tenacity of certain aspects of Hindu and Moslem culture" (Braithwaite, 1960, 822). Nevertheless, he judged that "although a certain amount of friction between Indian and Creole developed, and although there was a tendency to exploit the less sophisticated of the Indians, there was an over-all tolerance of the ethnic group and its culture" (Braithwaite, 1960, 828).

Studies of East Indians in Trinidad, carried out by U.S. anthropologists after racial voting became deeply entrenched in 1956 stressed the high degree of cultural retention and social encapsulation achieved in the island's rural communities. Klass, in his book on the village of Amity near Chaguanas (fig. 1) published in 1961, emphasized that in their life cycle, kinship, caste, and religion, East Indians had preserved important aspects of North Indian culture. The vast majority of indentured migrants were from the Gangetic plains between Delhi and Calcutta—that now constitute Uttar Pradesh and Bihar—with the main recruitment area centered on the villages and towns within a few hundred miles of Varanasi (Vertovec, 1992 and 2000).

In the racially homogeneous sugar belt in Trinidad, the East Indians' Hindu traits set them apart from Creole society. Nevertheless, Klass stressed that his village "is part of the ongoing social, economic and political system of Trinidad and cannot be legitimately separated from it" (Klass, 1961, 248). The Niehoffs reached similar conclusions in their examination of East Indian life on the Oropuche Lagoon, south of San Fernando (fig. 1), published the year before, and added that "the caste system appears to be less necessary for persistence of cultural identity than the Indian family type and hereditary religious beliefs" (Niehoff and Niehoff, 1960, 188).

In San Fernando, in contrast to the sugar areas, Indians were not only numerous, but they also lived in close proximity to browns and blacks. There, we proposed to explore the acculturation that had, by the independence period, taken place among Indians. We suspected that cultural differences—and racial exclusiveness—between Indians and Creoles had not been eradicated. We surmised that purely ancestral culture and social institutions among the Indians would have survived as ideals rather than as realities or endured only in name, but would have been creolized in character and function. Yet Indian culture and social organization, personality traits, and values were likely to be markedly unlike those of other Trinidadians—though to what extent San Fernando's East Indians differed from those in Trinidad's rural hinterland was for us to discover.

Geographical Context of the Journal

Out of Trinidad and Tobago's population of 830,000 in 1960, Creoles accounted for over 60 percent. The breakdown by color groups was as follows: whites, 2 percent; browns, 16 percent; and blacks, 43 percent. East Indians made up 37 percent of the island's population, among whom Hindus constituted 23 percent, Muslims, 6 percent, and Christians, 8 percent. Taking into account the neighboring island of Tobago (33,000 inhabitants) that had an almost entirely black population, Trinidad at independence comprised five racial zones (fig. 1): (1) The western sugar belt and its subsidiary rice-growing area, lying between Port of Spain and San Fernando, together with Naparima, to the south and east of San Fernando, and the sparsely populated southwest peninsula, were predominantly East Indian; (2) Port of Spain and its associated conurbation, stretching along the Eastern Main Road toward Arima, contained over 250,000 inhabitants, more than 90 percent of whom were Creole; (3) San Fernando, with a population of 40,000, was the only other town of note. Almost three-quarters of its population were Creole, and Indians formed the remainder; (4) the north and east of Trinidad were rural and sparsely populated, largely by blacks; and (5) by far the most racially heterogeneous locality was the Central Range, where mixed rural communities of blacks and Indians subsisted by small-scale cultivation supplemented by cocoa farming.

Although our journal describes visits made to most of the above regions, the greater part of it is dedicated to research experiences concentrated in the predominantly Creole town of San Fernando and the surrounding Indian sugar belt of Naparima. This region faces the Gulf of Paria (fig. 1), which separates Trinidad from Venezuela. In 1960, the Creole population of San Fernando comprised whites (3 percent), blacks (47 percent), and a mixed or brown component (21 percent), while the Indians were made up of three major religious groups: Hindu (7.6 percent), Muslim (5.8 percent), and Presbyterian (8.5 percent). As soon as we left the town by road, heading for Princes Town, Débé, or Penal, we entered the sugar belt of Naparima and moved from a Creole world into Hindu-dominated villages (fig. 1). Set up in the second half of the nineteenth century, these villages ensured that the former indentured laborers and their descendants stayed close to the cane fields and sugar factories that still required their seasonal labor.

San Fernando has an unusual topography. At the heart of the town stands Naparima Hill, around which the streets, running off an encircling road, reach in every direction (fig. 2). To the north of High Street lies the Spanish colonial grid, the east-west axis of which is paralleled by St. James Street, where we lived; to the south stands the promontory occupied by Harris Promenade, on which are located most of the important public buildings, such as the town hall, police station, courthouse, the Anglican and Catholic churches, and several schools.

At independence, whites occupied the high ridge at Spring Vale, to the north of the Spanish grid, and the enclave at St. Joseph

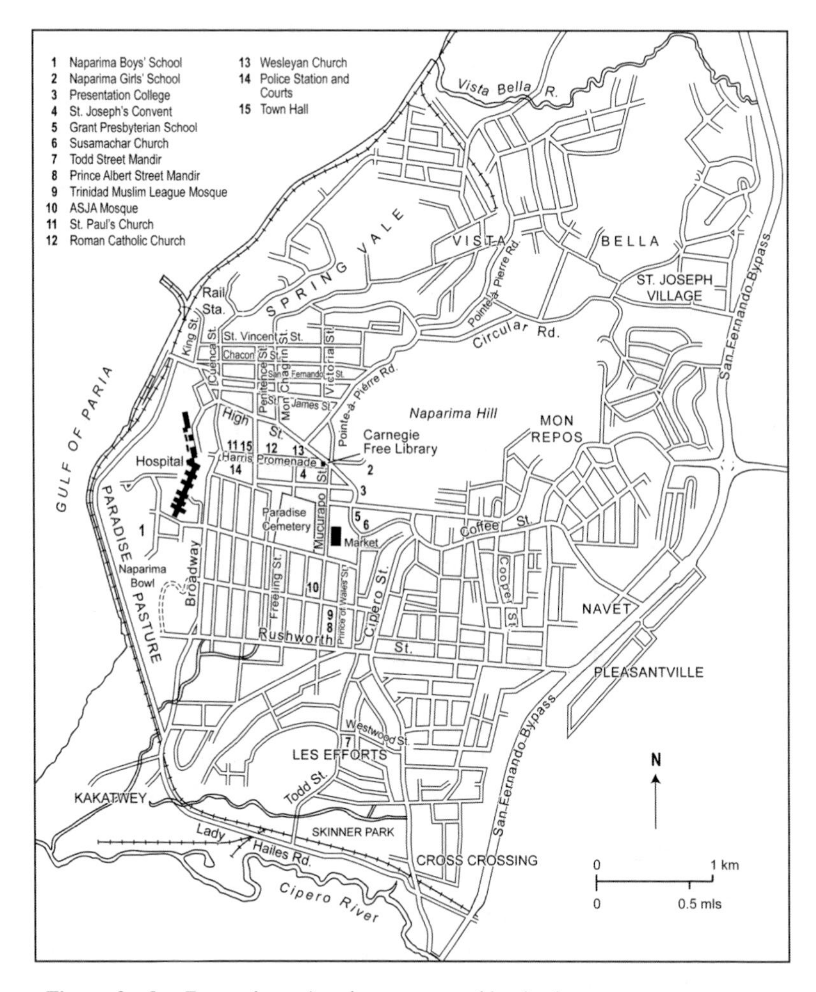

Figure 2 San Fernando: major place-names and institutions

Village, just inside the bypass in the northeast of the built-up area. Elsewhere, there was an intermixture of races and religions. The major exceptions were the government housing projects at Navet and Pleasantville, near the bypass, which were predominantly black and the Indian residential areas developed to the south of Paradise Cemetery. Here modest Muslim and Hindu concentrations focused on two mosques, and spilled south over Rushworth Street onto old cane land at Les Efforts, where the Hindu *mandir* (temple) was located.

The main road running due east out of San Fernando led to Princes Town, about six miles away (fig. 3). Off the road lay Ste. Madeleine, where the sole remaining sugar factory in Naparima (owned by Tate and Lyle) was located. There had been half a dozen *usines* (factories) in the vicinity in the early 1940s, but they had closed down because they were too small to operate efficiently. At Princes Town (formerly Iere Village), Rev. Dr. John Morton and his wife had founded the Canadian Presbyterian Mission in 1868. The missionaries extended their reach into San Fernando three years later and founded Susamachar Church (the Church of Glad Tidings) as the locus from which to convert Hindu and Muslim laborers and their families (Morton, 1916). It was because of the mission's establishment in San Fernando, and the inescapable fact that conversion led to urban, white-collar jobs, that Christian Indians in the colony gradually began to concentrate in the town.

Rural Presbyterians were vastly outnumbered by Hindus throughout Naparima, with particularly heavy concentrations of the latter located along the main road leading south to Débé, Penal, and Siparia (Richardson, 1975). A visitor to the swampy basin occupied by the Oropuche Lagoon described the village of Débé in the early 1920s as "almost wholly a Hindu town [*sic*], with a stream of many castes pouring down its highway," while neighboring Penal was noted for "its miles of Hindu vegetable gardens and its mud and reed huts that seem to have been transported direct from India" (Frank, 1923, 398).

Our Findings

We concurred with Klass and the Niehoffs that Indian culture, including religion, language, marriage, the family, and caste, though greatly simplified, persisted in Trinidad at independence; indeed, Hinduism and Hindi (technically Bhojpuri), Islam, and Urdu, all experienced revival after World War II as Hindu and Muslim leaders headed institutions that encouraged language classes held in rooms attached to their religious buildings. East Indians in San Fernando were an endogamous segment or, more accurately, a triad (Hindu, Muslim, and Christian) of religiously defined but linked ethnic groups external to the Creole stratification.

Christian East Indians were closer to Creoles in behavior and values than to their Hindu and Muslim counterparts, and Indian/black *douglas* were closer to Creoles than to their Indian progenitors. Although East Indian culture in San Fernando was being

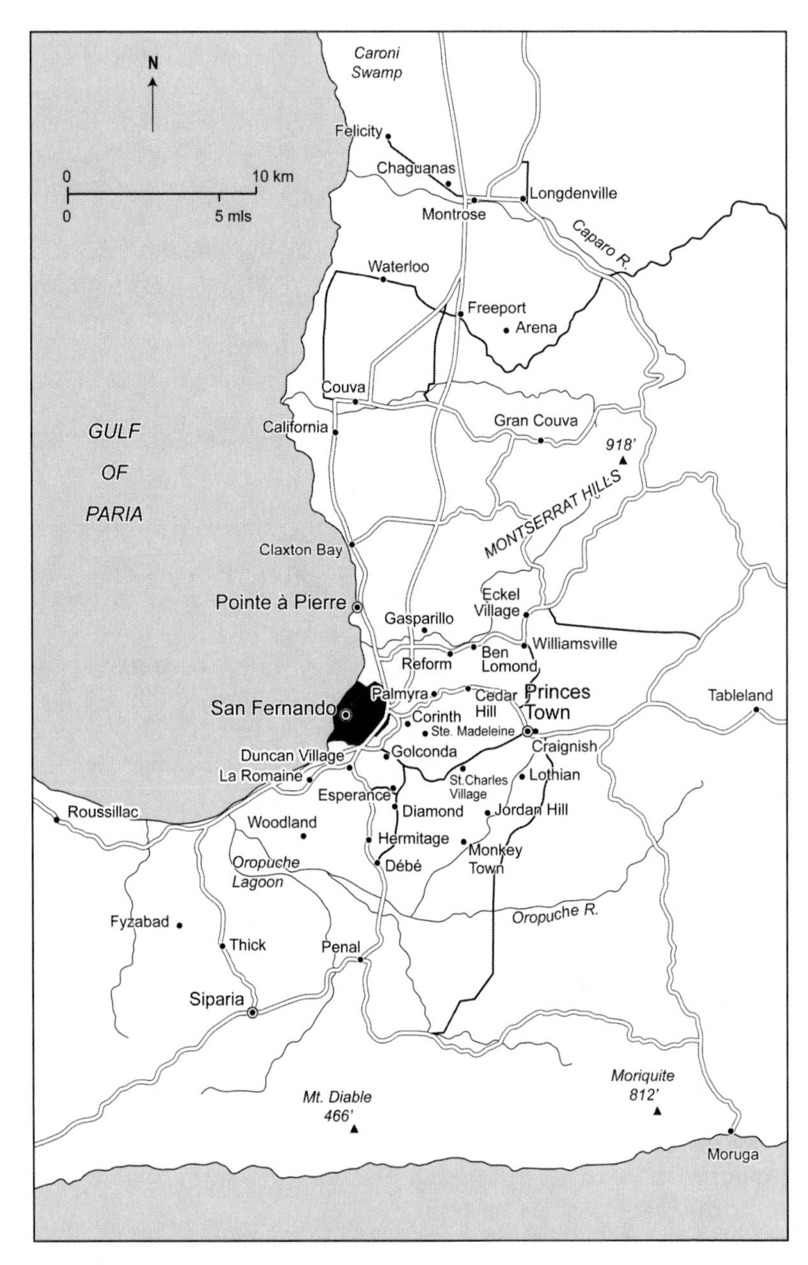

Figure 3 South-west Trinidad: place-names mentioned in the text, and roads

continuously renewed by the practice of finding brides from rural areas of Trinidad, and although Hinduism was further strengthened by reactionary Brahminism, urban Indians of all categories were more accommodating to Creole culture than were their rural counterparts, mainly through spatial proximity and social interaction (Clarke, 1986).

Christian East Indians (converted largely from the Hindu population in Trinidad) were partially acculturated to Creoles both attitudinally and behaviorally, yet linked to Hindus through kinship. In some cases, they behaved and believed as "Christian Hindus" (Rampersad, 2002, 12). Although Christian East Indians practiced Indian family systems and followed Indian dietary habits, they were marginally more likely to have married out of race than out of religion, and their friendships were more oriented toward Creoles than toward Hindus and Muslims. Against this, however, must be set the close kinship links between many Hindus and Christians, especially where conversion had occurred recently (Clarke, 1986).

Household data we collected showed that there was a clear, and to be expected, hierarchy of endogamy among San Fernando's Hindus. Almost half (49 percent) married within their caste; the majority (73 percent) married within *varna*—the larger caste category—and to other Hindus (86 percent); almost all had Indian spouses (Clarke, 1986). V. S. Naipaul confirmed: "marriage between unequal castes has only just ceased to cause trouble; marriage between Hindu and Muslim can still split a family; marriage outside race is unthinkable" (1962, 82).

Aisha Khan (2004) gives the impression that she was the first to identify creolization among Trinidad Indians, but our earlier research showed quite clearly that East Indians had been creolized by immigration, by indenture, and by the collective experience of more than a century of settlement in Trinidad. The traditional dress was rarely worn except on ritual occasions; in fact, it had been modified and more or less lost by 1964. As for language, Indians in San Fernando usually communicated with one another (and with others) in English Creole.

In the final analysis, however, it did not matter whether the Hinduism or Islam the Indians practiced was primordial or syncretic, provided that each was perceived to embody significant cultural differences. Despite considerable similarity in household composition between Creoles and Indians—based on the nuclear biological family (to which Munasinghe [2001] also alludes)—Hindu and Muslim hostility to mixed marriages ensured racial and cultural endogamy, thus guaranteeing the biological continuity of the East Indian segment (Clarke, 1986).

Urban East Indians have proven more receptive to Creole ways than their rural counterparts, and Christian East Indians have provided a bridge, however narrow, linking Indians and Creoles—both the Indian mayors of San Fernando in the decade after independence were Presbyterian and affiliated to the Creole-dominated political party. Indian-Creoles (usually black) or the *douglas*, were the outstanding candidates as go-betweens, though their numbers were small: culturally Creole, but with a residual orientation toward Indians that facilitated intermarriage and other social relations, they were potential brokers within the complex urban community.

Politics in the period of constitutional decolonization, starting with adult suffrage in 1946, transformed Creoles and East Indians into organized social groups—with generally negative consequences. The most outstanding example of this process was provided by the Sanathan Dharma Maha Sabha, an orthodox Hindu body. Its board of control transformed itself into the management committee of the People's Democratic Party (PDP) in 1952, thus extending Brahmin control from religion into politics (Malik, 1971; Ryan, 1972). So, when Dr. Eric Williams entered public life and formed the People's National Movement (PNM) in 1956, he was confronted by a racially and religiously defined opposition based on the Hindu population—first the PDP, and after 1958, when the PDP merged with other opposition elements, the Democratic Labour Party (DLP) (Malik, 1971).

Williams concentrated his efforts at the hustings on the black and brown Creoles, while making inroads into the Christian Indian and Muslim electorate and their leaders. It was through these informal coalitions involving Creoles and non-Hindu East Indians that the PNM won the two seats in San Fernando—and the national elections overall—in the 1956 and 1961 elections. Throughout our fieldwork, we found urban Creoles and rural Hindus locked in an intense and continuing struggle to control the government and resources of the newly independent state of Trinidad and Tobago and to define the national ethos of the country in the first years of its sovereignty (Malik, 1971; Ryan, 1972; Meighoo, 2003).

Political and Sociocultural Developments since 1964

During the forty-five years following our fieldwork, the significance of religious difference among the Indians has declined politically, and party affiliation has become even more determined by race. In 1964 no one could have predicted that the political hegemony of the Creole PNM would extend for an additional 22 uninterrupted years; Prime Minister Williams died in office in 1981 after a quarter of a century in power. Eventually, a rainbow coalition of Indians of all religious affiliations, disaffected middle-class Trinidad Creoles, and virtually the entire black electorate of Tobago formed the Alliance for National Reconstruction (ANR), and in 1986 consigned the PNM to the opposition benches (Ryan, 1988 and 1989).

The pattern that emerges may be described thus: Creole dominance through the PNM from 1956 to 1986, punctuated by black power riots in 1970 (Nicholls, 1971; Ryan and Stewart, 1995); a racial coalition under Robinson of ANR covering the years 1986–90 (Premdas, 1993), ended by a Black Muslim disturbance; East Indian government, through the United National Congress (UNC) in alliance with Tobago (1995–2002) (Ryan, 2003); and Creole dominance (1990–95) after 2002. The East Indian minority is able to form a government only through alliances with Creoles; Creoles may take electoral command of the state on their own, provided they do not lose control of part of "their" electorate, such as Tobago. However, neither of the state-threatening events that occurred in 1969 (black power) or 1990 (Black Muslim hijackings), involved the age-old Creole-Indian conflict. Mass discontent with the PNM in the 1980s provided the context for the ANR coalition; Creole loss of their "natural" dominance in 1995–2002 stimulated a reactive Creole nationalism that led to the electoral whittling away of the UNC majority in parliament.

The persistent and rarely punctuated involvement of race in politics, which goes back to 1956, or even to 1946, has had a substantial bearing on the national recognition given to some people who appear in the journal. A series of awards for outstanding contributions to the nation were introduced in 1969, replacing the British honors system. Focusing on Trinidad and Tobago's premier national—and most contentious—award, the Trinity Cross, which is given in recognition of "distinguished and outstanding service," we have mentioned in the text 11 people who were recognized by PNM governments: Rudranath Capildeo (1969) (interviewed), "Buzz" Butler (1970), Lord Learie Constantine (1971, posthumously), Rupert Archbald QC (1972), Lord David Pitt (1976), Dr. Wahid Ali (1977), Dr. Patrick Solomon (1978), Gerard Montano (1979), Mitra Sinanan (1979), Tajmool Hosein (1982), and Cecil Kelsick (1985). Two more were honored by the government of the Indian/Creole National Alliance for

Reconstruction, C. L. R. James (1987) and V. S. Naipaul (1989), both internationally recognized creative thinkers and writers.

More accessible to a wider range of those mentioned in the text is the second-ranking Chaconia Gold Medal, awarded for "long and meritorious service that promotes the national welfare or strengthens the community spirit." Professor Lloyd Braithwaite (1969), who is mentioned in this introduction, and Norman Girwar (1972), Sir Henry Pierre (1974), Dr. Ada Date-Camps (1981), and Mr. Justice Garvin Scott (1981) were recipients of this award from PNM governments between 1969 and 1981. Lionel Frank Seukeran (1985), Dr. Stella Abidh (1988) (interviewed), Alloy Lequay (1988) (interviewed), Professor George Sammy (1988, posthumously) (interviewed), Simboonath Capildeo (1989), Anna Mahase (1990) (interviewed), and Ashford Sinanan (1990) were rewarded by the Indian/Creole National Alliance for Reconstruction between 1985 and 1990. The record of these two awards, as reflected in the journal, suggests that the term "community" is conceived by Creoles as a black group into which only its affiliates and Indian allies are admitted; Indians without PNM approval are recognized only when their "community" has political leverage through participation in a national government.

The exclusion of East Indians from political power during the 1960s and 1970s (the Indian DLP did not contest the 1971 election) was a major factor in the Hindu revival in rural Trinidad, contrary to the expectations of most of our 1964 informants (Vertovec, 1992 and 2000). Hindu marriages, funded by rising wages in the booming oil and sugar industries, were transformed into massive, costly events (Ryan, 1991; Clarke 1993); individuals and communities vied with one another to perform expensive Hindu ceremonies; peasant Indian Trinidad was revitalized by a Hindu ritual culture with English as the main medium of communication. This cultural upsurge did not deliberately attempt to purify Hinduism or Islam of creolization (Khan, 2004). However, as the screening of Indian films waned, pirate videos took over, projecting modern Indian values and dress into Indian homes. The Hindu message to Creoles was loud and clear: you have political power, but we are different from, and superior to, you culturally (Ryan, 1999; Munasinghe, 2001).

Nonetheless, Hindu religious revival went hand in hand with creolization in many aspects of Indian life under the influence of the oil boom and the expansion of secondary and tertiary education. Revisiting Felicity in 1985, almost 30 years after his initial study,

Klass found that many of the East Indians had "adopted a Trinbagonian way of life—in language, in social and economic patterns, in material goods and aspirations. Even more, they shared with the rest of the island's population a commitment to Western education and egalitarian values" (Klass, 1991, 59).

Convergence with Creoles in values and taste (Miller, 1997) has harmonized with fusion in the arena of popular culture. Ryan notes that this "blending of the sitar and the steelband as well as the fusion of calypso, soca, and chutney music are now taking place in the musical hothouses of Trinidad and Tobago, and that the bowdlerised product now forms part of what constitutes an emergent 'neo-creole' or 'post-creole' aesthetic" (1999, 33). However, despite creolization or hybridization and the emergence of some "Indo-Afro-Saxons," Varadarajan (1992) insists that nothing fundamental has changed since independence, and that stasis is not entirely attributable to racial-political rivalry. The minority status of the Trinidad Indian in terms of demography "has made him [*sic*] defensive, inward looking, thwarted, endogamous, excessively protective of his culture, his food, his name, his rituals and religion, his women, his children, and his land" (Varadarajan, 1992, 13).

Our Positionality and Relationships

Gillian and I went to Trinidad in early 1964 as a British husband-and-wife team. I had graduated from Oxford with a BA in geography in 1960 and had already completed a year's doctoral fieldwork in Kingston, Jamaica in 1961, during which I was attached to the Institute of Social and Economic Research at the University College of the West Indies (since 1962 the University of the West Indies). While I was carrying out research on the changing social and urban structure of Kingston from slavery to the end of colonialism, Gillian was taking a BA degree in German with French at Bedford College, London University. She moved to Oxford University in 1962 and completed her Diploma in Education in 1963.

Our field research was guided by friendships with Bramadath Maharaj and Hansar Ramsamooj—our primary informants, both of them 20 years older than we were. In the journal text they are referred to, for simplicity, as Bram (Bramadath's nickname) and Hansar, though it was only on later visits to Trinidad that their first names replaced the more formal Mr. Maharaj and Mr. Ramsamooj. No one else we met, with the exception of Mr. (later Professor) George Sammy, and Mrs. Ena Scott-Jack, our landlady in San Fernando, offered such help, guidance, and comradeship as Bramadath and Hansar did throughout our fieldwork, though many others were courteous and informative during interviews.

The friendship with our two principal East Indian informants is difficult to explain; perhaps it involved some special affinity that the four of us—Gillian and I, Bramadath, and Hansar—had for one another; perhaps we offered them as much as they offered us. We became, however, aware of the need to break into new lines of enquiry toward the middle of the field research, and my interviews with the San Fernando elite, together with the contents of the questionnaire survey, provided counterweights to the valuable insights provided by our two Hindu friends.

Bramadath's strong sense of religious duty was an important factor in explaining his support for us. However, he accompanied us on many visits that involved non-Hindu activities, as well as Hindu ones, and at the end of our visit commented that he had, through us, learned a great deal more about Trinidad than he had previously known. In Hansar Ramsamooj's case, the issue was probably simpler. As a Hindu leader, he was keen on getting his modernizing, anticaste views across to us as researchers. Moreover, our interest in him and his religious movement undoubtedly bolstered his position in the Hindu community, in which he was appreciated for his religious message "you get in life what you focus on," yet marginalized because he was not a Brahmin.

George Sammy provided us with several crucial interviews about his experiences of growing up in Trinidad, and about East Indians in general and Madrasis in particular. He made contact with many people whom we later interviewed, and organized and hosted the focus group in the middle of our research, which guided the questions we selected for use in our questionnaire. In addition, he read a draft of the questionnaire for us, and commented both on its content and on the appropriateness of the "English" used to elicit answers.

Ena Scott-Jack introduced us to her family and friends, and facilitated contacts for us throughout the Creole community in both San Fernando and Port of Spain. She ironed out many of the practical difficulties of daily living in Trinidad from nonvalid visas to car accidents, from night time intruders to con men. Unfailingly cheerful and sociable, she ensured that we had a more relaxing and wide-ranging experience of Trinidad than the research schedule strictly permitted.

One asset we had in conducting our fieldwork was the research I had already carried out in a Creole urban community in the Caribbean,

Kingston, Jamaica, where I had kept an academic journal. In addition, Gillian's degree in German had given her a fine sense of language, an interest in writing text, and a trained memory for detail. After a few weeks we found that we could reproduce scenes and conversations in considerable textual detail. However, while we can claim with some authority that the journal is not about us, we cannot deny that it is about Indians and Creoles in San Fernando and the adjacent area, filtered through us. In short, while we are writing as sympathetic and interested outsiders, *we* are always selecting where to go, with whom to speak and along what lines, and what to note down.

As a couple we had access to both genders, though readers will note the virtual absence of women from the record of prominent interviewees. However, we did have discussions with Dr. Stella Abidh and Anna Mahase, arguably the most prominent Indian women professionals of their generations. Many of the conversations recorded by Gillian with Indian women excluded me and most of the conversations I had with Muslim men and male leaders of the Creole and Indian communities, in their own homes or business places, were recorded while Gillian was checking the progress of the questionnaire survey. Without two scribes, it would have been impossible to maintain the journal entries on such a regular basis or at such length, especially during the last two months of the fieldwork, when we were at full stretch.

More than 45 years on, we have both retired—Gillian from Wycombe Abbey School, a leading English academic school for girls, where she was Head of German and Careers, and I from Oxford University, where, as an Emeritus Professor of Geography, I am still involved with the graduate program. In retirement, Gillian has continued to sit as a magistrate in the Adult and Youth Courts in Oxford. On reflection, we consider that our questioning of respondents would, today, be more acute and penetrating than it was in 1964, but we are by no means sure that Trinidadians would have opened themselves up as completely to our older and more established selves as they did then.

Text of the Journal

The text we have written does not argue its way analytically through an issue, as in a research paper; instead, each entry adds layers of meaning, in the same way that an artist adds oil paint to a canvas. The narrative has been copied verbatim from our handwritten documents, kept in three notebooks (with ancillary information on loose sheets of paper), but cut by about 20 percent, at the request of the publisher, to remove repetition and trivia.

Much of the material in the first half of the book, covering the months January to June, was written up in full at the time from memory or from notes kept during ceremonies or events we attended. The detailed interviews in the second half, covering July to September, were recorded on the spot in note form, and have been reworked as a narrative for this publication. In a few instances, sentences or paragraphs have been reordered to consolidate themes or to improve comprehension, and minor adjustments have been made to the text to improve the style, avoid repetition, clarify meaning, or avoid giving offense. To assist the reader, the text has been divided into three parts: Settling in, Taking Soundings, and Conversations.

Rather than interrupt the text with supporting or ancillary material, endnotes are provided at the culmination of the part in which they are cited. The footnotes fulfill several functions. They explain the English meaning of the Hindi (Bhojpuri) or Urdu terms (see also the glossary); they also specify the book or article in which an issue may be pursued further. Some of the material in the footnotes relates to publications that have appeared since the fieldwork was carried out, and thus contextualize our journal entries. Footnotes are also used to comment on the accuracy of the material in the journal text.

Finally, footnotes also provide biographical details about the journal's more important interviewees. Two main sources of information are used: the 1945 "Who's Who" of leading East Indians, published as a section of the *Indian Centenary Review, One Hundred Years of Progress, 1845-1945, Trinidad, B.W.I* (Murli J. Kirpalani *et al.*, 1945); and *Who's Who in Trinidad and Tobago, 1966* (Carlton Comma, ed., 1966). The former, which appeared 19 years before the research began, and the latter (the first national *Who's Who*), issued 2 years after the fieldwork was completed, contain information on 29 East Indians (for 1945) and 23 East Indians, and 17 Creoles (for 1966) who appear in the pages of the journal. Seven East Indians are common to both the 1945 and 1966 lists.

The language of the journal is Standard English, with phrases from time to time in English Creole (Indo-Trinidadians prefer "coolie English"; see glossary) to capture a pithy remark by one of our interlocutors. Trinidad English Creole is spoken in a register that, unlike Jamaican Creole, is readily accessible to Standard English speakers, though some of our interviewees also use the odd Spanish and French Creole word, explanation of which is provided in the glossary and footnotes. Catholic sugar planters from the French Caribbean had been encouraged to settle in Spanish Trinidad with their slaves after 1783 (14 years before the British conquest), and in the early nineteenth century, "French Creole (locally called Patois) was still considered by many to be the lingua franca of the island" (Winer, 2009, xiv).

Most Trinidad Creoles and Indians, other than members of the upper classes, now speak English Creole (or "coolie English") all the time, but older Indians, at the time of our fieldwork, were fluent in Bhojpuri (the language of Bihar and the adjacent parts of Uttar Pradesh, though originally treated, mistakenly, as a variant of Hindi). Bhojpuri or Urdu (for Muslims) was widely used in the 1960s in conversations in Trinidad's rural villages, and in Hindu and Muslim rituals in urban and rural areas. During Hindu and Muslim ceremonies we would usually be given a running commentary in English, and it is these, together with key Bhojpuri and Urdu terms (see glossary) that are recorded in the journal. Occasionally, English versions of songs or prayer sheets were provided for participants.

Conclusion

During the last two months of our field research, many interviewees, especially members of the San Fernando elite, referred to others whom we had interviewed earlier. Through these linkages and through the substance of our conversations, the complex webs of race, religion, and politics have been woven into the text of the journal. It seems to us that the issues of cultural plurality or segmentation (and of hybridity) in Trinidad and Tobago are not so much debated as palpable. Yet, it is important to underline that most members of the San Fernando elite insisted that, in their urban community, pluralism was negotiable, not doomed.

We hope that readers will enjoy our journey of enquiry as much as we did in 1964, and will appreciate the intensity of the interviews (and the research in general) in the last few weeks, when we were at full stretch to get words down on paper. Propelled by the velocity of our research activity, the text develops a momentum not unlike the plot of a novel.

When we left Trinidad, we traveled from Ena Scott-Jack's house in San Fernando to stay with George Sammy in St. Augustine. The next morning Bramadath Maharaj and Hansar Rasamooj collected us and took us to the airport at Piarco. These four friends not only supported us throughout the fieldwork, but also provided us with hospitality on return visits to Trinidad. We dedicate our book to their memories with warmest thanks.

Notes

1. Trinidad (794,624 population) and the neighboring island of Tobago (33,333) taken together had a population of 827,957 in 1960. Tobago was linked to the colony of Trinidad by imperial fiat in 1889.
2. A plural society is racially and/or culturally complex, and has usually had a colonial history of minority dominance by whites over the majority of nonwhites who may constitute ranked or parallel groups.
3. Trinidad and Tobago's colonial parliament (the Legislative Council) was unicameral and had nominated as well as official members in addition to those who were elected on a restricted franchise prior to 1946, and by adult suffrage thereafter. For the last colonial election in 1961, in anticipation of independence, the title of the legislature was changed to the House of Representatives, and all the seats were filled by election. The title Member of the Legislative Council was superseded by Member of the House of Representatives, and a second chamber was created in the shape of a nominated Upper House, whose members are called Senators.

Part 1

Settling In

Tuesday, January 28, 1964

Gillian: *Arrival, Port of Spain*

Having enjoyed our visits to Jamaica, Puerto Rico, and Antigua, Colin and I were almost reluctant to reach Trinidad last night. Here it is hotter and, at first sight, less beautiful than our previous destinations. There is a bustle and a quickening of pace in Port of Spain (fig. 1). We are looking forward to settling in San Fernando and beginning work in earnest. Jack Harewood, the director of the Government's Central Statistical Office, will, I think, be most helpful. A quiet, gentle man, he has shown us great kindness today in driving us to San Fernando (fig. 1) to try to find somewhere to live. We have located possible accommodation in two-thirds of a fine wooden house. Unfortunately it is expensive and in need of a coat of paint.

Port of Spain seems full of East Indians,[1] at least where we have been—the airport, taxis, wharves, and boarding houses. Around Queen's Park Savannah, the elegant houses are mainly wooden with fretwork. Port of Spain is cleaner and more attractive than Kingston. Our boarding house, at 96 Maraval Road, a traditional wooden house, slightly decrepit and with few mod cons, is run by Mrs. Edna Roxburgh, a formidable lady who, despite a recent heart attack, seems full of energy.

I can tell from his quietness that Colin is strained. We must just struggle to keep our heads above water for the next few days. How can we come to grips with San Fernando? It is situated on the side of a much-quarried hill. The roads are steep and tortuous, and the houses packed tightly together. But it is attractive, in a higgledy-piggledy way, and the few people we have met there were most friendly. I must just mention that Vera Rubin's[2] predictions were correct. No sooner had we stepped into the taxi from the airport last night than our East Indian driver began to complain about Dr. Eric Williams, the Creole prime minister and leader of the People's National Movement (PNM).

Wednesday, January 29

Gillian: *San Fernando, 67 St. James Street*

We now have a home in San Fernando. The grey-painted wooden house we saw yesterday at 67 St. James Street, just below the junction with the Pointe-à-Pierre Road, is to be our base during our stay in Trinidad (fig. 2). We decided to make our own way to San Fernando today to make further enquiries about accommodation, and, after a bumpy ride on the Express bus, we managed to contact Mrs. Ena Scott-Jack, who owns the house. She is black, early middle-aged and very helpful. Because of the contact with Mrs. Roxburgh, she has agreed to reduce the rent from BWI$175 (£35) per month to BWI$150 (£30).[3]

We are satisfied with the house. Standing on concrete stilts, beneath which Scott's Commercial School is located, the main construction is of wood, with a fretwork surround where walls and ceiling meet. We have a large living-dining area, two bedrooms, a study, kitchen, shower and toilet—enough for a family. The location in the town seems good, in that we are fairly central and not in a conspicuously wealthy area. I think that our neighbors opposite are Hindu, judging by the faded *jhandi* (prayer flags—see glossary) in the front garden.

San Fernando is hard to take in at first. Streets wind here and there, up and down and around. Houses are perched one on top of the other on the slopes leading down to the sea from Naparima Hill. Most are on wooden, stone, or concrete stilts, which means that there is a living space beneath the house; almost all look dilapidated and frail. The shopping centre along High Street is untidy, the gutters littered with garbage, banana and orange peel. Pungent aromas pervade the gloomy food shops. Poorly dressed men and women squat on the pavement trying to sell their wares—corn-on-the-cob, fruit, vegetables, and jewellery. Ragged beggars hold out imploring hands. And all day long, the sun burns fiercely down on the town.

The rural areas between Port of Spain and San Fernando are, as one would expect, predominantly East Indian. Most of the land is under sugar and rice. Hindu temples and Muslim mosques are to be seen in all the major settlements along the road. It is worth noting that the houses are, for the most part, larger and better quality than comparable rural homes in Jamaica, and they are almost all raised well above the ground—high enough to park a car underneath, to house tools or a pottery.

Thursday, January 30

Gillian: *Kitchener's tent, Port of Spain*

We had a busy day in Port of Spain, the high spot being the visit to Lord Kitchener's[4] tent. The calypso "tents," an important part of

Carnival, bear no resemblance to tents as we know them. All the stages in the town are taken over, and each is dominated by an established calypsonian such as Lord Kitchener or the Mighty Sparrow. There is a band (in this case a magnificent one led by Frankie Francis), which accompanies all the calypsonians as they perform. We heard a range of performers, including Calypso Rose from Tobago (the only female performer), King Fighter from British Guiana, and Zebra from Grenada.

Apart from Lord Kitchener, who has clearly made the big time, they are all very ordinary people. But Kitchener was disappointing. None of the songs we heard were as good as our Sparrow record with "May May" and "Benwood Dick." Kitchener's songs were either colorless or, in the case of "The Well," crude. On the whole, the smutty calypsos didn't compare with Sparrow's because they lacked wit, apart perhaps from Lord Conqueror's song about the cockfight.

In sharp contrast, Lord Pretender was severe, straight-faced and unashamedly didactic. His songs addressed race (his point being that we are all members of the human race) and the recent hurricane disaster in Tobago. He, too, was an older man and, like all the others, black. We saw no East Indian calypsonian last night. Apart from a moving tribute to the late President Kennedy,[5] the calypsos were not particularly topical, nor was there much social comment, apart from digs at Sparrow, (who has stood down from the Calypso King contest this year), and some mocking of East Indian marriage customs, such as returning the daughter to her parents if she is not satisfactory.

The décor in the theater captured the spirit of carnival with masks, huge sprays of paper flowers and brilliant, non-garish colors, and there was something delightfully homespun about the whole show. As far as the audience was concerned, the high spot was Cowboy Jack with his yodeling—not a calypsonian!

Friday, January 31

Gillian: *Mrs. Roxburgh's version of a Hindu wedding*

Mrs. Roxburgh described for us, in some detail, a recent Hindu wedding in Port of Spain, concerning the wealthy Jang Bahadoorsingh[6] family, who live opposite her in a large house on the Maraval Road. The first son had left the island and married a white girl, but the second (aged 21) married an Indian girl, aged about 16–17. The Hindu premarriage celebrations lasted for approximately one week at the boy's home, with much singing, beating of drums, and festivities late into the night.

On the morning of the wedding, there was a celebration for about 600 people at the groom's house with vegetarian Indian dishes and alcoholic drinks. According to Mrs. Roxburgh, few guests would attend if there was no alcohol at all. Then everyone drove off to the bride's parents' home, where the main ceremony with 1,500–2,000 guests was to take place. The bride wore a yellow sari with matching jewellery, while the groom was in a white suit with tight trousers and a Nehru-style coat, buttoned from top to bottom. The ceremony was performed with a ring made from the sinew taken from the back of a bamboo leaf. Once more the food was Indian, and again alcoholic drinks were available.[7] Then the couple and guests returned to the groom's home for a western-style ceremony, during which the bride now wore white and received a diamond ring. An interesting amalgam of eastern and western traditions!

Sparrow's tent, Port of Spain

We visited the Mighty Sparrow's[8] tent in the evening. This was quite different from Kitchener's, which seemed sophisticated in comparison; an open-air hall with a canopy roof, a rough wooden ramp for a stage and masses of bright, bold adverts everywhere. The adverts were so dominant in this tent that the calypsonians looked more like Punch and Judy figures as, visible only from the waist up, they nodded, shook, and wiggled behind the barrier of cardboard. We heard many calypsonians this evening, most of whom sang only once. I have recorded their names: Robin, Starling, Mighty Clipper (East Indian), Mighty Dougla, Breego, Mighty Killer, Duke, Lord Laro, Striker (Calypso King 1958, 1959), Composer (part East Indian), Lord Blackhat, Scaramouche, Mystery, Caruso, Cristo, Fish (a strange comedian from Barbados—we couldn't understand a word), Rambler (part East Indian), Beaver, Lord Nelson, Lord and Lady Fluke.

The whole show was less pretentious and more homespun than Kitchener's. I had no idea last night just how slick his show was. Sparrow himself was disappointing. Where are his wit, his sauce, his clever variations of sound? He just stood in front of the microphone and shouted. But the audience still loves him. No one can deny his mastery as an entertainer. He's powerful, vital, explosive, but all this fails to add up. His road march had a haunting tune, but his second song, "The Village Ram," in which he paraded his sexual prowess, was crude.

However, if "Sparrow day done," the future looks good for some of his fledglings. Laro, an amateur calypsonian, charmed the entire audience with "Don't Buy"—a topical song dealing with the boycotting of Johnson's store on Frederick Street. Starling, a close follower of Sparrow in his early days, sang a delightfully saucy number about teaching someone to twist.[9] Lord Nelson, another in the Sparrow mold, was disappointing. Duke, an erstwhile southern Calypso King, was also good. These three and many others outshone the Mighty Sparrow.

On the whole, the songs were more topical and moralistic than last night. Again we had the deification of the martyred Kennedy. Race, Kennedy's assassination, and religion were frequent themes this year. Social commentary centered on the picketing of the Frederick Street store and the Lock Joint sewerage scheme. One singer claimed that the Government should concentrate on getting enough food for everybody before providing the means to get rid of it! There was some reference to different customs in Trinidad and some gentle mocking of "coolies."

Saturday, February 1

Gillian: *Maracas Bay*

On Saturday morning, we drove northeast out of Port of Spain to Maracas Bay through the beautiful Northern Range, heavily wooded with lush tropical vegetation and patches of red immortelle. There was an Indian settlement behind the beach at Maracas, with wretched wooden shacks.

Steel band

In the evening we went with Dorothy, an elderly American friend of Mrs. Roxburgh, to hear some of the Silver Stars Band practicing in a back yard. This band consists mainly of college boys from Queen's Royal College and Queen Mary College. They had gathered in a disused, rubble-covered area behind the houses, not far from where we are staying. Groups of people were milling around enjoying the music. The tone of the band was mellower and less strident than some we have heard, though they all sound better out of doors. The boys, who are strictly loyal to one band, rather like a gang, make and tune their own "pans" from old oil drums. A carnival band with all its nonmusical performers can be 1500 strong.

Sunday, February 2

Gillian: *Savannah, orchid show and the Hilton*

Another full and exciting day spent in Port of Spain, with a morning walk along the Maraval Road, around Queen's Park Savannah and through St. Clair, mainly looking at different architectural styles. At worst they can be too heavy, like some Victorian municipal buildings in England, but I love the ornate wooden houses with their delicate jalousies, fretwork, and verandas.

Then we went by taxi to an orchid show at the Hilton Hotel. I had no idea how beautiful and varied these plants are. Most of the people there were light or white, and all seemed to know each other. The Hilton itself is a surprise. It is known locally as "the upside-down Hilton" because the entrance is on the top floor, with the bedrooms built into the hillside below. Well frequented by the people of Port of Spain, it gives panoramic views both into the hills, and away over the bay and the Gulf of Paria toward Venezuela, and down to San Fernando's denuded hill.

Steel band contest

After lunch, we walked in the blistering sun to the Savannah to hear the preliminary round of the steel band contest. With difficulty, we found somewhere to sit in the stand so that we were in a good position to hear each band going past. Some bands, mostly no more than about fifty strong on this occasion, had their pans painted in their own colors. All were mounted on wheels and were played on the move. Among the players were people, mostly young boys, whose job it was to push or pull the pans along, in order to allow the bandsmen to concentrate on playing the tune. The Silver Stars were undoubtedly the scruffiest, though they would rightly claim that it is not what they look like that counts! My favorite band was Starlift, a small group of youngish boys.

Throughout, the crowd was jostling, munching, drinking, with groups of fans rising to their feet to applaud their favorite band as it processed before the judges. But my abiding memory is of the insistent drumming of the combined bands. Some made a harsh metallic sound, when wheel hubs were struck with bits of steel. The pans are beaten with rubber-covered sticks. The most popular tunes were "Mama, dis is mas," "Kokiako," and "Come leh we go." Some of the bands were so close together that the tunes overlapped.

Few East Indians were to be seen playing in the bands, and they were notably scarce in the crowd, though there were one or two young Indian boys helping to push the drum racks along in the bands. Silver Stars is known as the fair boys' band; they were certainly lighter on the whole than most other bands, though some were dark; one was East Indian and one white. This can be attributed to the fact that Silver Stars is a college affair, unlike most other bands, which are based on trade groups.

Tuesday, February 18

Gillian: *Settling in, San Fernando*

There has been a gap of more than a fortnight, in which nothing has been recorded, though much has been achieved. We have moved into 67 St. James Street, which is distinctly bare inside, as we have hardly any personal belongings of our own; and we have bought a 1958 Volkswagen Beetle, which has made life much easier.

The house has brought plenty of worries. It was the base of the Norwegian Mission to Seamen, but has lain empty for some months. In a tropical environment, this means that the insect world has had ample time to establish itself. Cockroaches have colonized all the built-in, wooden drawers and cupboards, concealing themselves behind the woodwork where they can be smelled but not reached. In the evening, mice scamper around the living room and kitchen. Remedies have been sought for both nuisances, and now we hope there will be some improvement. Mosquitoes nibbled us nightly until we invested in a costly, though invaluable, mosquito net.

From my point of view, things are fairly hard. I had intended cooking West-Indian dishes from the start, but soon found that it takes me all my time to cook in any fashion. Food is expensive here and the quality poor. It is easy to spend a fortune on roach killer and mice poison before ever getting round to buying something to eat. The housekeeping isn't simple. The atmosphere is dust-laden, largely because of the many roadworks in our area, caused by Lock Joint's program of sewer installation. Uncarpeted expanses of dark wooden floor need dusting and polishing, and we have what feels like a lot of rooms to keep clean. But we are managing. A woman from the town does some washing for us at 30 cents an item, so we have to limit the amount we send. I find myself trying to juggle domesticity and helping with the research. Between us, we are just about managing to keep things moving, despite the heat, which is draining.

Prelude to Carnival, San Fernando

Carnival, the high spot of the year here, was upon us in no time. Ena, our landlady, has been immeasurably kind; without her, we could well have had a lonely couple of days. On the Saturday before Carnival (February 8), she took us to a "jump-up" (dance) at the Promenade Tennis Club in San Fernando. This is a solidly middle-class (or even upper-class) club. Fortunately, we were made welcome. Ena, having dropped us off, soon left for another party in Port of Spain, so we were looked after by Mr. Donawar and his daughter, Muriel,[10] the defeated PNM candidate at Fyzabad in 1961. We all had a wonderful evening dancing, or, in our case, stepping or shuffling to the calypso music. There were a surprising number of East Indians at the club that night. Ena said that all barriers are dropped during Carnival, only to be reerected on Ash Wednesday.

On the Thursday (February 6) before Carnival, two days before the Promenade Tennis Club dance, we had gone to the Queen and King of the Bands Show at the Naparima Bowl. This is a natural amphitheater on the far side of the Princess Margaret Hospital in San Fernando. The show was enjoyable but long; it lasted from 8 p.m. until midnight, and it was packed. Most of the white and lighter-colored people were sitting in the covered seats, while the rest, including us, sat in the bleachers.

Some of the men's costumes were stunning—notably Julius Caesar, Richard Lionheart, and an Indian Chief. The finest craftsmanship had gone into making dazzling headdresses from materials such as brass, gold, and silver. In contrast, the band was poor. It must surely have ranked among the weakest in Trinidad. Lord Kitchener, Rose, and Mighty Bitterbush struggled to sing to the feeble accompaniment.

J'ouvert (the opening day, or Carnival Monday, February 10) in San Fernando turned out to be a gloomy affair. Soon after dawn, people lined the pavements in the town center, waiting for the bands to pass. As there are effectively no tourists in San Fernando, Colin and I stuck out a mile. Time and again we were menaced by the oil-covered *jab molassies* (molassies devils). I found them sinister and threatening.

By far the best part of *J'ouvert* was waking up to the throb of the steel bands as they rolled into town. No recording can capture the beauty and haunting quality of the steel band on the street. Again, at about 9 a.m., we paused in the middle of breakfast to throw open the doors and windows in order to fill the house with the powerful surging of the bands on their way home. But this time, the players were beating out hymn tunes, and the followers were singing too. It was an enchanted moment, the air filled with music and the sky with light, with the rust and salmon buildings of San Fernando glowing in the freshness of morning.

Carnival, Port of Spain

But we fled to Port of Spain, to the heart of the celebrations, where we regained our anonymity and blended with the tourists. Much of the day was subdued, though it livened up in the evening with a "jump-up" in Diego Martin. Here we danced till we dropped, on a crowded, ramshackle wooden dance floor set up in a private garden. My main memories are of aching legs, perpetual thirst, and despair of finding somewhere to sit. There were several English people at this party. I'm thinking in particular of a UN town planner, who sniffed contemptuously at our proposed research focusing on the East Indians, admitting, at the same time, that it would probably be a matter worth looking into. The solicitor-general[11] sneered. His brother, on the other hand, Eric Kelsick, was both interested and helpful.

We spent the night, or what remained of it, in Cynthia Scott-Patrick's house. Cynthia is Ena's sister, though she seems different from her siblings. Ena and their brother Garvin (Mr. Justice Garvin Scott[12]) are both extroverts, whereas Cynthia is quieter and more subdued. I think she is probably a widow, as she has grown-up children, all living away from home. At night the house is guarded by two dogs; to my horror, they snap at our heels every time we go along raised wooden decking to the outside lavatory. Heaven help a would-be intruder!

Tuesday (*Mardi Gras*), was the big day. We sat for hours in our seats in the grandstand, glutting ourselves on the magnificence on parade. There are so many gorgeous costumes that it is impossible to remain equally enthusiastic about each one. After a while, the audience clapped only the extraordinary confections. They all deserved claps and cheers, but as a crowd we were only bestowing our appreciation on those with the grander costumes.

The best part had nothing to do with the major display; it was hearing the bands in the morning passing Cynthia's house as they went along Tragarete Road. That was indescribable but quintessentially Trinidadian. In the evening, Ena and her friend, Royal, took us along to "jump-up" with Starlift. There was already an element of weariness in the air as the band beat its way through the streets of Woodbrook. By now the older masqueraders had dropped out and the younger ones had changed bands in order to be with friends. Some were in full costume, others in half costume, and the rest like us in trousers and bright shirts. As far as I can remember, there were no East Indians "jumping up" with us.

First steps in San Fernando

After Carnival, a period of preparatory work began for us in San Fernando. Hours have been spent going through newspapers and cutting out relevant articles. Ena has introduced us to several people, among whom was an Indian couple. But we have a lot of reading and writing to do. Church on Sunday (February 16) was an experience, but we found Canon Farquhar[13] mildly irritating. The service took place at six o'clock in the morning and the church was well attended. All Canon Farquhar did was to talk down to us as if we were children, chiding us with our disregard for Lent. He is light-colored, elderly, with grey hair and an impeccable English accent. Moreover, he is the father of Peter Farquhar,[14] the Liberal Member of the House of Representatives for Pointe-à-Pierre.

The most markedly East Indian areas around San Fernando are the rural areas of Débé and Penal (fig. 3). A few days ago, when we visited them, we were arrested by the young Indians we saw—lovely dark-eyed, raven-haired children, mostly looking very poor. Indian houses tend to be reasonably big and raised on tall stilts. Underneath the house there is invariably a hammock, which immediately calls to mind Sam Selvon's[15] Trinidad novels with Urmilla rocking her child.

So far Indians are proving elusive; nearly all our contacts are black. Norman Girwar,[16] who was suggested to us in Kingston by Pundit Tewari, is the only East Indian we know. The Cane Farmers' Conference in Montrose (near Chaguanas), to which he invited us, was Indian dominated. Of the 65 people attending the conference, only 15 were black. The proceedings closed with the Lord's Prayer—a compromise, perhaps? Mr. Ramyan Ali was fasting for Ramadan. Although Indians seem friendly once contact is made, actually making that initial contact is not so easy.

Wednesday, February 19

Gillian: *Magistrates' Court, Moruga with Will Hercules*

After an early start, Ena, Colin, and I went with Will Hercules to the Magistrates' Court at Moruga. Will was to defend a mother and son in a coffee-stealing case. More precisely, the pair were accused of having stolen coffee in their possession. The drive down was unexpectedly beautiful, first through rolling fields of cane and later through deciduous forest with mile upon mile of flaming immortelle.

Moruga is a village on the south coast, with a magnificent shoreline. The main street leads down to the sea, ending abruptly in a rough grassy patch before the sand begins. On the seafront stand a Roman Catholic Church and school. The police station, courthouse, and shops are clustered together at the seaward end of the street. It would seem that a fair bit of fishing goes on there, since several boats were drawn up on the sand, and one fisherman was mending his nets. There is no harbor as such; Moruga simply nestles in the middle of a long bay. To left and right of the village rise wooded cliffs. Inland, behind Moruga, there is said to be much distilling of "bush" rum.

The court was a long room, with the officials and defendants at the front, and rows of benches for the public and defendants waiting for their case to be called. At the beginning of the morning session, there were approximately 65 people in court. At the most there cannot have been more than 20 Indians. In order to maximize ventilation, all windows and doors were wide open. Unfortunately, this meant that the sound was dispersed, making it difficult for us, squeezed together as we were on a bench at the back, to hear everything that went on.

We gathered that all those with cases pending were called before the magistrate and either dealt with on the spot or had their case adjourned. There were many adjournments and few completed cases. Most involved petty crime, such as using obscene language or larceny. One man was charged with beating and causing harm to a dog, another with beating another man. The accused in the latter case, an East Indian, was drunk when giving evidence, which caused some mirth in the public gallery. The courtroom was packed throughout the morning, and during the afternoon until the coffee-stealing case, clearly the case of the day, was finally dismissed. Will had reason to be pleased with his performance.

Some of the East Indians in the court looked desperately poor. Most people, however, had put on their Sunday best, and were clearly enjoying a day out. All the women had their heads covered, even if it meant wearing a crumpled little hankie. The fact that a surprising number of defendants failed to appear for their case caused little consternation and appeared to be a regular occurrence.

The local rum shop opposite the Court House had an excellent day, for the majority of those not actually involved in a case, and without a seat inside, spent their day drinking. As soon as the magistrate wanted a particular defendant or witness, he would call out the name, which was relayed to the two policemen at the door. These then shouted the name over to the rum shop, and depending on the response, reported whether that person was present or not!

Great care was taken over the swearing-in of witnesses, each denomination or religion being given an opportunity to take the oath in the appropriate manner. Anglicans and Methodists swore on the authorized version of the Bible, while Catholics preferred the Douet version; Muslims swore on the Koran and Hindus on a brass pot or *lotah*. Devout Chinese are said to take a saucer and smash it.

Lunch with David Lee Chow

Will took us into the local store, a rambling emporium owned by David Lee Chow, who, unusually, has two wives. The first is Chinese. He married her in China and she lives with him and their children in Moruga; the second, a Creole, lives in Port of Spain with their children. During the school holidays, the Creole wife moves to Moruga and installs herself plus children in the family home, while the Chinese wife moves temporarily into a small house in the village. This is, by all accounts, an amicable arrangement! Lee Chow gave us a delicious lunch consisting of many small dishes, simply served. In Will's opinion, Lee Chow serves the best Chinese food in Trinidad.

Thursday, February 20

Gillian: *Oropuche Lagoon and Siparia*

We had a late afternoon drive to Siparia through gently rolling and at times wooded countryside. From the Oropuche Lagoon on, the whole area becomes markedly East Indian, apart from Creole pockets in the former oil region of Fyzabad. This predominantly Indian area is very different from that around Caroni, and at first sight appears more prosperous. In the clearings, there is a great deal of sugar cultivation, but we saw only one example of rice-growing, in Thick. Once more the countryside was dominated by the vivid, flame-colored blossoms of the immortelle. Just as you enter Siparia on the main San Fernando road (not the Thick one) you can see mile upon mile of immortelle growing in the forest.

The main aim of our visit was to see the famous black virgin, La Divina Pastora, in the Roman Catholic church of the same name at Siparia. This is an effigy of the Virgin Mary, which mysteriously, either due to the corrosion of the paint on her face, or through a miracle, has turned black, or, to be more precise, dark brown. The

Niehoffs[17] write at length about this in their book. We found the virgin, unexpectedly, in a fine new church. Her effigy is small, and she was exquisitely dressed in a white wedding dress with a dainty coronet and veil on her head. Around her neck, there hung countless strings of pearls and gold chains.

Monday, February 24

Gillian: *Inadequacies of the English*

I went to church with Ena this morning while Colin did a land-use survey of High Street and the surrounding area. After the Anglican service, Ena introduced me to Canon Farquhar. I groaned inwardly when he said, "I'm always particularly pleased to see English people in church." Imagine my surprise when he continued, "They really are such pagans!"

Last Saturday, when Hari Lal had finished polishing the floor of our house with the hand polisher, a long heavy brush on the end of a stick, we suggested that the best way to keep the polish up, would be for us to go over the floors several times a week. He looked surprised and said that we would have to borrow Mrs. Jack's electric polisher because the alternative would be too hard for Colin! And this from a tiny, skinny little chap, who looks as if a puff of wind would blow him away. But he clearly believes that people like us are just not strong enough to do our own manual work. I must add that we are unusual here in that we cook, clean, and wash the car ourselves. Moreover, I also do most of the washing myself, by hand. St. James Street is a modest sort of area, but most families here employ a maid.

Wednesday, February 26

Gillian: *Indian houses*

Many East Indians, particularly those in the Caroni and Oropuche areas, own large concrete houses (Plate 1). Both wooden and concrete houses are constructed to the same design on stilts. Even the older wooden ones in the Oropuche Lagoon have cars parked under the house. The new, concrete houses—the expensive and the poor—are mostly identical. The significant difference between them is that those belonging to poorer people are rarely finished; the bricks are raw and the house soon resembles the shack it succeeded.

Plate 1 An *ajoupa* with thatched roof, wooden house, and modern concrete houses on stilts, edge of the Oropuche Lagoon, Débé

Thursday, February 27

Colin: *Reflections on Indians met*

Today is the East Indian celebration of Holi or Phagwa,[18] which marks the beginning of the Hindu year (a day of secular fun celebrated in Trinidad as Indian Carnival). It seems an opportune moment to look back on the East Indians we have met during the past month. On our arrival at the airport in January, we met an East Indian taxi-driver named Singh, who took us into Port of Spain. He was, in his own words, "very alive to the racial situation." He clearly saw all political and social issues structured in terms of East Indian and Creole. He bitterly attacked the Lock Joint sewerage scheme and the government's failure to hold three by-elections in areas held by the PNM. He was also convinced that entry into the Civil Service was difficult, if not impossible, for East Indians.

Singh was extremely proud of Vat 19—a Trinidad rum, but scathing about the blacks and small-islanders living in Port of Spain's Shanty Town (East Indians live there too). He was determined to condemn Eric Williams on any imaginable ground. I told Singh that I had been informed by a Canadian we had met in Antigua that many

North American businessmen suspected Williams of being communist. Singh misunderstood me and thought that I had said anticommunist. Nevertheless, he began an all-out attack on Williams for his anticommunism. I pointed out his mistake and Singh then swung into an equally violent attack on Williams as a communist. Singh thought that things were going to get so bad that he was preparing to go with his wife to England or Scotland.

A source of local information relating to East Indians is Eric Kelsick. "I am not a Trinidadian, so you can take my views as those of an unbiased observer. Surely, most of the East Indians here must have come from the lower castes in India. The absence of a really good Indian restaurant is a clear indication."

According to Eric, racial pride among East Indians is comparatively recent and goes back only five or six years. Before that, they were pleased to marry blacks. This no longer holds true. People are proud of being Indian, and sophisticated (acculturated) Indian women in Port of Spain attend cocktail parties wearing saris. I noted one East Indian woman in a sari collecting goods at the Hi-Lo supermarket in San Fernando.

Eileen and Clayton Appleton, friends of Ena's, are both highly creolized—to the extent of eating *souse* (pigs' trotters) and spending their leave in England. Clayton, whose father was a *dougla*[19] (black-East Indian mixture), said that Eileen was East Indian—but he was West Indian. Ena has two other part East-Indian friends, Girlie Gomes and Grace Kangaloo (née Namsoo). Girlie is half Madeiran (Portuguese) and half East Indian; Grace is three-quarters East Indian—her mother's mother was from Scotland. Both are proud of their East Indian forebears, as are the Appletons. Girlie claims to know a lot about East Indians, but the other three know little.

Holi

Today has been the Hindu festival of Holi. This morning I asked Mr. Ragounath of Marabella Bay Road whether there would be any celebrations in the streets of San Fernando. He replied that street celebrations were not allowed by the police. He is going to celebrate in Penal tomorrow. People form bands there and sprinkle one another with *abir* (a pink dye).

Gandhi ashram, Todd Street

This evening I went to the Hindu *siwala* (Shiva temple) in San Fernando. The watchman, a Muslim, showed me around the building. The hall or *ashram* is modern and large enough to seat 60–100 people. A service is held at 8.30 a.m. on Sundays. The altar area, or *mandir*, has a large wooden dais in front of it, and in the middle, a stone *lingam* (phallus) surrounded by petals. The *lingam* is painted green, except for the top portion. A special container hangs over the *lingam*, allowing water to drip steadily on to it, drop by drop. Two pottery crucibles (*dias*) were burning on the altar. There were several Hindu statues (*murti*) set against tiles, and the back wall was decorated with illuminated pictures of Krishna and other gods. The figure of Hanuman, I later discovered, was made in Couva; that of Krishna was a gift from India.

I went to a house behind the temple to meet the *saddhu* (holy man or mendicant) and was invited to approach by a girl of about 14 or 15, who called to her *bap* (father). The *saddhu* was about 60 years old and spoke in a way that suggested that Hindi was his mother tongue. I remembered the greeting "*Namaste*," which we exchanged. He said he would be pleased if we went to the service on Sunday.

I asked about the festival of Holi. He said the Indians would be celebrating it in India and British Guiana, but not in San Fernando. "The businessmen here are too busy making money to celebrate," he said rather bitterly. His daughter added that there would be no celebrating on the streets here, but both agreed that there would be celebrations in Penal and the country parts. The difference between Creole and East Indian festivals is summed up here. For Carnival one goes to town (San Fernando or Port of Spain); for Holi one goes to the country.

I asked some East Indian girls at Scott's Commercial School (located in the area under our house) whether they knew anything about Holi. None did, although one was Hindu. But most knew when the teacher referred to it as "Indian Carnival." All agreed that it was not celebrated in the streets of San Fernando. Very little is known in Creole circles about East Indian festivals, but the same is true the other way round. The Indians cash in on Carnival as tailors and sellers of dry goods.

Saturday, February 29

Colin: *Mr. Dialdas, businessman*

I went to see Mr. Dialdas (another contact provided by Pundit Tewari) in his shop on High Street. He sells Indian dry goods, wedding dresses and saris as well as *dhotis* (loincloths). His East Indian wife was wearing a sari. Mr. Dialdas was born in Sind, India, and went to Jamaica in about 1932. In 1938 he came to San Fernando, and, because of the outbreak of war, stayed on.

Mr. Dialdas has two children, a boy and a girl, both at college in San Fernando. He says that his father taught him his business, but his own formal education was limited. He speaks Hindi better than he speaks English. With the conflict between Hindu and Muslim at India's partition in 1947, Mr. Dialdas's family had to leave Sind and go to Bombay. They lost their store and lands. This has been a lesson, and Mr. Dialdas hopes that his two children will get a good education. After completing School Certificate, they may go to India for training.

Mr. Dialdas has been back to India and plans to go again this summer. His brother went in his place last time because he had received a promise of marriage. (The brother has a dry goods shop on Frederick Street in Port of Spain.) Mr. Dialdas sells saris and white European wedding dresses to East Indian brides. The sari is usually of a saffron color and is worn for the Hindu wedding ceremony. The white bridal gown is put on after the binding part of the ceremony. Needless to say, this is not done in India. It appears that in Trinidad a small but increasing number of brides are wearing the sari throughout the wedding.

Mr. Dialdas was the chairman of the committee that built the *mandir* and hall (Gandhi Ashram) and erected the Gandhi Memorial on Harris Promenade. (This statue of Gandhi stands close to that of Columbus, the "discoverer" of Trinidad.) Mr. Dialdas now seems to have little to do with the Hindu *mandir*, which is run by less wealthy Hindus. He supplies costumes for both Carnival and the Ram Lila (Rama) pageant: his preoccupation is with remaining neutral. Several of the storekeepers on High Street were born in India—Mr. Dialdas, Mr. Kirpalani[20] (Bombay), and another from Gujarat.

While I was in the shop, I was offered a soft drink, and later introduced to Sarbjeet Pundit and his wife Sookdayia. They invited me to visit them at Cedar Hill early next week.

Notes

1. Port of Spain is essentially a Creole settlement, as Stephanie Goodenough shows in "Race, Status and Residence, Port of Spain, Trinidad," Ph.D. thesis, University of Liverpool, 1976. However, East Indians do concentrate in jobs associated with transport and hotel services, as the text indicates.

2. Dr. Vera Rubin, Director of the Research Institute for the Study of Man, New York, whose institute funded our research project.
3. In 1964 one British West Indian dollar was worth four shillings UK (equivalent to 20p in decimalized currency).
4. Lord Kitchener (Aldwin Roberts), one of the leading calypsonians of the late colonial/early independence period. He was both Calypso King and multiple winner of the annual Road March for the calypso most popular with the steel bands.
5. President Kennedy was assassinated in November of the previous year—1963.
6. Jang Bahadoorsingh, born 1913, manufacturer, merchant, and Hindu leader ("Who's Who," Murli J. Kirpalani *et al.*, eds., *Indian Centenary Review: One Hundred Years of Progress, 1845–1945,* 133).
7. This account does not conform to our experience of Hindu or Muslim weddings, though alcohol, and allegedly meat, is sometimes available in a household adjacent to the one in which the ceremony is being held.
8. The Mighty Sparrow (Slinger Francisco)—often referred to simply as Sparrow—was the dominant figure in Trinidad calypso in the years before and following independence.
9. The "twist" was a dance popular in the United States and the United Kingdom in the late 1950s and early 1960s.
10. Muriel Donawa-McDavidsen, born 1929. Muslim. Lost Fyzabad in 1961 to Vernon Jamadar, DLP, by 126 seats. PNM Member of the House of Representatives 1966–91; Minister with various portfolios in the Williams and Chambers PNM governments from the mid-1970s to the mid-1980s. Died 2001.
11. Cecil Arthur Kelsick, born Montserrat 1920; educated Montserrat Grammar School and King's College, London University; called to the Bar, Inner Temple, 1941. Holder of various senior legal positions in the Windward and Leeward Islands 1950–57; Solicitor-general, Trinidad and Tobago, 1957– (Carlton Comma (ed.), *Who's Who in Trinidad and Tobago, 1966*, 143). Chief Justice, Trinidad and Tobago, 1983–85. Trinity Cross, 1985.
12. Mr. Justice Garvin Scott, born San Fernando, 1916; educated Naparima College, San Fernando and Queen's Royal College, Port of Spain; called to the Bar, Middle Temple, London, 1945. Magistrate, 1951; Crown Counsel 1954; outstanding legal service in West Africa 1954–62; Judge of the High Court of Justice, Trinidad and Tobago, 1962– (Carlton Comma (ed.), *op. cit.*, 216). Chaconia Medal, Gold, 1981.
13. Canon Max Farquhar, born 1894. Educated Grammar School, Freetown, Sierra Leone and Codrington College, Barbados; ordained priest 1920. Rector St. Paul's Church, San Fernando, 1950– (Carlton Comma (ed.), *op. cit.*, 95).

14. Peter Farquhar, son of Max Farquhar, born 1927. Educated Queen's Royal College, Port of Spain and the London School of Economics. Accountant and economist; Industrial Relations Officer, Shell Trinidad Ltd. Elected DLP Member of the House of Representatives, Pointe-à-Pierre in 1961; 1963 defected and helped to form the Liberal Party (Carlton Comma (ed.), *op. cit.*, 95).
15. Samuel Selvon, *A Brighter Sun,* 1952; and *Turn Again Tiger*, 1958.
16. (Surendranath) Norman Girwar, born 1915; educated Naparima College, San Fernando and Government Training College for Teachers. Solicitor and conveyancer in San Fernando from 1950; Manager, Trinidad Islandwide Cane Farmers' Association 1958–. Member, San Fernando Borough Council 1953–56 (Carlton Comma (ed.), *op. cit.*, 104). Recipient of the Chaconia Medal, Gold, 1972.
17. Arthur and Juanita Niehoff, *East Indians in the West Indies*, 1960, 155–6.
18. Since 1967, Holi or Phagwa has been reinvigorated as a national celebration, having previously fallen into decay (Selwyn Ryan *The Jhandi and the Cross: The Clash of Cultures in Post-Creole Trinidad and Tobago,* 1999, 171).
19. *Dougla* (pronounced doogla) is the Hindi for hybrid.
20. Murlidhar J. Kirpalani, born Hyderabad, India, 1907. Came to Trinidad in 1927 and established Kirpalani United Company. Chairman, India Famine Relief Committee. Pioneer in introduction of Indian films to West Indies. Lives in Port of Spain. (Kirpalani *et al.* (eds.), *Indian Centenary Review*, 145).

Part 2

Taking Soundings

Sunday, March 1

Colin: *Sandhya puja, Hindu temple*

At 8.30 a.m. we went to the Hindu temple on Todd Street (Plate 2). The sandhya puja (daily prayers) had not yet started, as Mr. Ramsamooj was still en route from Corinth, a village on the eastern outskirts of San Fernando, where he conducts an early morning service. Mr. Bramadath Maharaj introduced himself, saying that the *saddhu* had told him that we were expected. We were given a service sheet for the *puja*, or ceremony, and the words of several *mantras* (ritual utterances). Once Mr. Ramsamooj had arrived, the worship proceeded with hymns and prayers, sung and recited in unison, in Hindi and English alternately. Mr. Ramsamooj led the singing. At midpoint, five boys and girls went up to the altar one by one and took a *dia*, moving it in a clockwise direction and performing *aarti* (worship) before the images of the gods and over the blue *lingam*. The number five is believed to be propitious by Hindus.

Plate 2 Gandhi Ashram and, under the spire, the *mandir* (temple), Todd Street, San Fernando, looking south toward the cane-fields of Naparima

Mr. Ramsamooj then addressed the congregation of 40–50. He was brief and to the point, going over the meaning of two or three verses in the first *mantra*. He concluded by welcoming Gillian and me to the service. Another *mantra* was sung and the congregation dispersed at 10 a.m. As they left, they took *persad* (sacred food), consisting of dried prunes, coconut, raisins, sugar and bananas from a wooden tray handed around by the *saddhu*. Using the third finger of her right hand, a barefoot girl placed a sandalwood *tika* (a spot indicating religious devotion) in the center of everyone's brow—the third eye. Another little girl carried a *tarriah*, or round brass tray, with *dias*, in which the collection was placed. Each donor put two hands over the brass tray and the lamps and raised them to their heads.

We were introduced to several members of the congregation. They were extremely friendly, and one man, who works for the Sugar Welfare Board, offered to take us into the rural areas. He introduced himself as "Bisram Gopie, OBE."[1] The older women all wore the *oronhi*[2] (head veil), the younger girls, white mantillas.

Monday, March 2

Colin: *Gita Class—Mr. Bhattacharya*

At six o'clock in the evening we went to the temple to hear a reading from, and explanation of, the Bhagavad-Gita conducted by Mr. B. Bhattacharya.[3] He is the Principal of Tagore College at Craignish, Princes Town, a nondenominational private secondary school. Mr. Bhattacharya read from the Gita in Hindi, explained in English the shades of meaning of every significant word, and drew philosophical parallels with other world religions, especially Christianity. He is knowledgeable about the New Testament and rightly assumed that his East Indian audience (about 60 strong) is too. One must remember that all adult East Indians in the South of Trinidad, who have had any schooling, received it at the hands of the Canadian Mission.

The standard of Mr. Bhattacharya's scholarship is patently high. He has just started a second class at Débé. There, he says, his scriptural analysis is less intense. The men are full of admiration for Mr. Bhattacharya—the pundit from India, and the fact that they have been able to put up with the labored pace of six chapters in 18 months indicates their genuine interest in the subject and their philosophical turn of mind.

Mr. Bhattacharya is a reform movement in himself, purifying and Sanskritizing the peasant Hindu religion. He aims to give it philosophical status and a sophistication that can enable its adherents to cope with the challenge of Christianity. Mr. Bhattacharya is at pains to indicate the rifts and dogmas associated with Christianity and to argue that certain basic principles underlie all the sects of Christianity—and that these same principles are part and parcel of the teaching of the Bhagavad-Gita.

Tuesday, March 3

Colin: *Sarbjeet Pundit, ex-indentured laborer*

At 2.15 p.m. we set off for the home of Sarbjeet Pundit at Cedar Hill, in the Princes Town direction, which we found with some difficulty. The pundit was sitting at a table on the veranda of his small wooden house when we arrived. We were welcomed by him and his wife and were offered seats in their sitting room. The room was bare. The walls were of wood, as was the unpolished but carefully swept floor. (The house was raised on low stilts, which must have been quite high at the back where the ridge dropped down to the cane fields.)

The sitting room contained a settee, an easy chair, a rocking chair, and one or two upright dining chairs. The walls were covered with brightly colored pictures of Rama and Krishna, Ganesh, and calendars showing gods in similar scenes. The calendars are distributed by prominent East Indian firms; framed pictures of religious importance for Hindus are sold in the arcade and on street stalls in San Fernando. The sole piece of modern equipment was a radio, though we understood that the pundit's son has a motor car.

I asked the pundit to tell me his name. He went into the adjoining bedroom and brought me his certificate to practice as a priest. It was dated 1937, and was signed by the then colonial secretary. Neither the pundit nor his wife can read or write English, though both can read Hindi and write Devanagari script.

Sarbjeet Pundit was born in a village about 30 miles south of Gorakhpur in the United Provinces of India. In 1911, when he was 21, Sarbjeet went on a pilgrimage to Benares (Kasi, Varanasi). He was met by a recruiting agent, who promised that he would be able to make a lot of money if he signed up and went to Trinidad as an indentured laborer. He agreed, and when he landed in Trinidad, he was sent to the sugar plantation at Cedar Hill, where he worked in the cane for four years. Once his indentureship was over, Sarbjeet started working as a pundit, and cultivated a plot of land at Penal (while living at Cedar Hill), growing coffee, cocoa, and sugar. During this period he married Sookdayia, now his wife. She lived in Penal with her parents until her wedding. Sookdayia's father was born in India: his home was Benares (Varanasi or Kasi), and he too, she claims, was tricked into enlisting for Trinidad with the tale that a fortune could be made there.

Sarbjeet Pundit's household comprises him and his wife, his son and pregnant daughter-in-law, and their five children. The daughter-in-law is fair-skinned (as is Sookdayia), and quiet. Her children are beautiful, and her youngest son is clearly a favorite with his grandparents.

Sarbjeet Pundit was wearing a *dhoti* and *kurta* (Indian shirt)—but no *janeo* (sacred thread).[4] His wife wore the *oronhi*. They are both strict vegetarians. It was interesting to see that, despite their seemingly poor circumstances, they did have an Indian servant. She spoke Hindi, too, and had tattoos in Devanagari on her forearm. Before leaving, we were each given *persad* from their prayer room.

Thursday, March 5

Colin: *Visit of Bramadath Maharaj, Hansar Ramsamooj and others*

By 7 p.m. we were ready to welcome our two East Indian guests—Bramadath Maharaj and Hansar Ramsamooj. To our surprise, our guests numbered eight—not two. Mr. Ramsamooj brought his daughter Rosalind, Miss Rampat, Miss Hari Lal, Miss Ramkissoon, Mr. Sookhoo from Princes Town, and Mr. Treboransingh from Gasparillo. Gillian was given a bouquet of pink roses and anthuriums.

I asked about East Indian participation in Carnival. Mr. Sookhoo claimed, and the others agreed, that most East Indians who are still Hindu do not take part in the festival. For them it is essentially black and Catholic. Acculturated Indians, especially the youngsters, "jump-up," but virtually none "play mas." Trebu added that he stayed at home. All agreed that there was great danger of racial conflict at Carnival and that the safest policy was to keep out of the way.[5] Several times during the evening this tactic of avoidance was mentioned. Our guests also agreed that Carnival should not really be regarded as a national festival; to call it a Creole festival would be more appropriate.

Mr. Sookhoo is a teacher in the Presbyterian School in Princes Town. He was educated at a Canadian Mission School and at

Naparima Training College. He is rather unusual in that, despite this close contact with church schooling, he is still a Hindu. Indeed, he is the only Hindu on the staff of the Princes Town Canadian Mission School.

Mr. Ramsamooj (Hansar) and Mr. Maharaj (Bram) both pointed out that the two main political parties are run on strictly racial lines, with no appeal to policy or effectiveness whatsoever. Both feel that although Eric Williams is nonracist, his speeches must have racist undertones to satisfy his followers. For example, "Massa Day Done"[6] has implications for Indian employers of black labor.

Bram is a supporter of the Democratic Labour Party (DLP). After the last election, he was relieved that the DLP was not the government because he wondered how this would be accepted by the civil service and blacks generally. Would strikes have brought down the new government, as in British Guiana? All agree that it is better to play second fiddle and live quietly in their own way. This is another example of avoidance. Both Bram and Hansar say it is a good thing for the black to be top dog here at last—and for a while at least—but that all races should be accorded the same respect.

Bram says that race, employment, and promotion are closely linked—even at Texaco, where the policy of management is racially neutral. Blacks get rapid promotion to supervisory jobs because they are prepared to bully the men into activity. Hansar is an electrician at Texaco, Pointe-à-Pierre, and Bram works in the stores at the Texaco research laboratory, also at Point-à-Pierre.

Hansar has 11 children, ranging from the ages of twenty-plus down to three. Bram, who is estranged from his wife, is the father of three boys and three girls, aged between 18 and 11. The eldest boy has several GCE 'O' Level passes—the equivalent of a grade 2 Cambridge. He is an apprentice electrician in the oilfields. The remaining five children are all at secondary school. The East Indians are keenly aware of the importance of education if they are to improve their competitive position in this society.

Both Hansar and Bram are enthusiastic supporters of cricket. (Hansar's brother plays as a professional for Northamptonshire in the United Kingdom). Both note the absence of East Indians from the Trinidad side—and of southerners generally.

Although older people can remember where they and their parents were born, they seem to have little knowledge generally of the source area from which the Indians were drawn. Mr. Sookhoo, for example, thought that they were mostly from Madras in south India.

Saturday, March 7

Gillian: *Puja at Hermitage*

The following is a description of an evening *puja* at Hermitage, just outside San Fernando in the cane area off the road to Débé. Hansar and his group, the Selfless Service Divine Mission (SSDM), regularly go out into the country to hold house or village *pujas*. These prayer meetings are usually held on a Saturday night, though the form of service is designed for the early morning, noon or evening. The latter is the only time available to the SSDM group, as it best suits the workers who are out in the cane all day.

In this instance we were to witness a household *puja*. On arrival, we found that the yard in front of the house on stilts had been transformed into a makeshift temple for the occasion. Rows of benches were set out before the *vedi* (altar), and blankets and old sacks for the children to sit on had been spread over the ground, which was tarred, no doubt with pitch from La Brea. Many villagers and relatives had gathered for the occasion, and a large number of children were milling around. As far as I can remember, only one of the local women and one of the daughters from the family involved in the *puja* wore saris. All but one of the young women from SSDM wore white saris.

Before the *puja* started, the devotees—host family—were nowhere to be seen, but villagers and friends had been gathering for some time, and they all chatted excitedly to each other in both Hindi and English. It was mostly the older people who spoke Hindi with each other. Some of the older Indian women looked worn and poorly dressed. Their clothes had clearly been washed over and over again, and in several cases were beginning to fall apart. Their faces were weather-beaten and deeply lined, their features frequently sunken. They were nearly all thin and tiny.

The actual *puja* began much later than planned. Once everyone was finally ready, Hansar, in *dhoti* and *kurta*, called the Mohun family down from their house. They all appeared dressed in white, the father in *dhoti* and ordinary shirt, and the mother in a ballerina-length dress with *oronhi*. The elder daughter wore a white sari and veil, while two young sons, much to their own and their friends' amusement, wore white vests with a homemade *dhoti*, pinned together from a muslin-like material, shaped like a large nappy.

The *vedi* was decorated with the symbol of the sun. The other two symbols that can be used are those of the cross and the swastika. Any color may be employed, but on this occasion, the rice covering the surface of the *vedi* had been painted to produce a mosaic effect in red, black, and gray. The center of the sun was gray, its projecting rays, red, and the background, black. On the altar stood *dias*, which were kept alight throughout the service. The dishes and jars used by the devotees during the ceremony were made of brass, not of clay. The devotees washed their hands with water lifted from the *lotah* (brass jar) on a mango leaf, though a banana leaf could equally well have been used.

The ritual of the *puja* was sufficiently complicated for the devotees to rely heavily on the directions given them by Hansar's assistant, Bisram Bissoon. Bisram also made the necessary preparations for the offering of rice, *ghee,* and camphor, which was mixed in a broad-bottomed brass dish and ladled onto the fire by all members of the family. The male head of the family and the assistant together made the offering of the young coconut at the end of the *puja*.

After the final offering had been made, Hansar, wearing the wooden beads and pink tassel denoting a vegetarian, explained the meaning of the service to the assembled gathering. He stressed that the offering of material goods on the altar and the cleansing of the body by the devotees was only symbolic. It was not an end in itself, but a means of preparing the individual for communion with God. Camphor, he continued, was used to light the fire because it leaves no ash; likewise, each individual should burn his life with the purity of camphor. The act of fanning the fire symbolized the life of the individual and the significance of God for that life; fanning the fire meant keeping God alive within the human heart. Finally, he explained the sacrifice of the young coconut as symbolizing an offering to God of a humble object that was nevertheless part of God's creation.

When the ceremony was over, the host family handed to each member of the assembled gathering a small brown paper bag containing *persad*. This is a Hindu sacrament, made from flour and *ghee*, sugar boiled into a syrup, and perhaps some ground rice. Hansar found some fig (banana) in his packet, which he rejected in no uncertain manner.

After the service, Hansar and Bram led us upstairs into the house, where we were offered a meal. In fact, only Hansar, Colin, and I ate. Bram acted as chief waiter and intermediary between the host and us. For example, he made detailed enquiries about the content of the food in order to make sure that it was strictly vegetarian (for Hansar's benefit). On being questioned by Bram whether the food really was vegetarian, Mr. Mohun, the host, replied emphatically, "Strictly vegetarian. Perhaps not on other days, but today strictly vegetarian." From his searching questions, it was clear that Bram was far from convinced. He ate nothing himself, but hovered around advising us what and how to eat.

Cold water from a jug was poured over our hands to wash them, Hansar blessed the food and we began to eat with our fingers—a novel experience for us. Bram confessed that he feels embarrassed when he has to use cutlery in a restaurant. We drank pineapple juice and ate *roti*, rice, lettuce, tomatoes, pumpkin mayonnaise, hot pepper mayonnaise, and pea mayonnaise. The pepper and pea were particularly spicy, but we enjoyed the food, and it was a memorable experience to be eating in a rural Indian home. Mr. Mohun's brother's daughter, who had come over from Penal for the occasion, helped serve the food.

Although Mr. Mohun referred to Hansar as pundit, this is not technically correct. He is what one might describe as a lay pundit, performing many of the duties of a pundit without actually being one. As a non-Brahmin, he is excluded from the ranks of the traditional pundits.

Monday, March 9

Gillian: *Hindu women visitors*

Last Thursday evening (March 5), as Colin has already recorded, we were visited by a group from the Todd Street *mandir*. There were four young women among the gathering, two of whom are seamstresses, and two who remain at home to help parents. None of them are married, though at least two must be in their late twenties; all work at or from home. Apart from the general discussion, we "women" had a long talk. They were trying to encourage me to wear a sari, claiming that Indians regard it as a compliment when westerners wear it. They themselves were wearing western dress, but all have saris for special occasions, notably religious functions.

Miss Harilal told me something about her family's religious life. They are practicing Hindus and have a special prayer room in the house. Six o'clock in the evening is prayer time for Hindus, and at that time, incense is burnt before the gods. Miss Harilal says that the incense candle is always lit in her home before any other light in the house is put on. In answer to tentative enquiries about the significance of caste to them, they all emphasized that the most important thing is to be a good Hindu rather than to be of high caste.

Miss Harilal is clearly intrigued by our small families and use of birth control. When I confirmed their suspicions that we have no children, despite having been married for 18 months, Miss Harilal leaned over to Miss Ramsamooj and said, "All they does have to control theyselves!" I don't know whether she meant self-control or birth control, but there was a note of great sympathy in her voice.

Second Gita class

This evening (March 9) we attended our second class by Mr. Bhattacharya on the Bhagavad-Gita. Again we sat and listened for a good two hours. He is an intelligent man, a devout Hindu, and a gift from God (or rather the gods) for the Hindus here in Trinidad. While the exposition of the Gita is fascinating, his criticism of Christianity is at times unconvincing. But he knows how to hold an audience, and his talks are stimulating. It is remarkable how he continues to draw so earnest and keen a class from the ranks of ordinary people, many of whom, like Hansar and Bram, have received little or no formal education, but have no difficulty in following and understanding these philosophical talks.

For us, Mr. Bhattacharya's textual analyses provide an insight into Hinduism today, though at times I find myself, to my surprise, wanting to jump to the defense of Christianity. The talks provoke reflection on Christianity in the context of other world religions. All in all, I would rather listen to Hansar's short simple sermon than to Mr. Bhattacharya's lengthy philosophical dissertations. Hansar is self-effacing, whereas Mr. Bhattacharya is self-confident. But Mr. Bhattacharya's skill as an orator is considerable. His speech is poetic, the imagery well chosen and beautiful. Dressed all in white, he sits cross-legged and barefoot on a table draped with fabric, wearing a *dhoti*, an elaborate *kurta,* and a fine silk scarf.

During the lecture, the *saddhu* comes in and "cleans up" the temple gods, that is to say, he removes the day's faded flowers and lights the *ghee* in the *dias.* He performs *aarti* by moving a lighted *dia* several times in a clockwise direction in front of each god. Once this is completed, he replaces each *dia* in front of the appropriate god. Before leaving, the *saddhu* prays in front of the blue, marrow-like *lingam.* He then bows his head, touching his thumb and fingertips alternately with his nose.

After the lecture, we met two Christian Indians, Mr. George Sammy[7] and Mr. Peter Dubé, both of whom regularly attend the Gita class. Mr. Sammy is a workmate of Bram; both before and after getting his BSc. from London University, Mr. Sammy used to work in the research laboratory at Texaco. He has recently registered for a doctorate and has been appointed lecturer in Chemical Engineering at the University of the West Indies at St. Augustine.

Thursday, March 12

Gillian: *Frank Cleghorn's visit*

This morning brought a surprise visit from Frank Cleghorn of Penal. We had first met him and his wife, Myrtle, at the Promenade Tennis Club "jump-up" on the Saturday before Carnival. They are both East Indian, have five young children, and knew the Niehoffs when they were researching in Penal. Frank is a government sanitary inspector, but outside his working hours, he is also helping with the Agricultural Census. He and his wife are Christians. She spends much of her spare time doing work on behalf of the Canadian Mission Church in Penal.

Frank accepted a glass of rum, though he later revealed that his favorite drink is whisky. He is a keen meat eater, with a penchant for a good beefsteak. He explained something of the sanitary conditions in rural areas. At present he is particularly concerned with the incidence of hookworm, which is passed from one person to another by treading on human feces. The worm attaches itself to the skin of the foot before making its way into the intestines. The problem cannot be solved simply by providing lavatories or holes in the ground. Frank told us, for example, of a *saddhu* for whom a new toilet was built. The old man was duly grateful, but he locked the door and announced that it would remain unused because he preferred the cane field.

Bram

Bram dropped by in the evening to make arrangements for Saturday's *puja*. We spoke a lot about Hinduism, and the observation, which sticks most in my mind, is that he is sure that Hinduism in Trinidad is doomed. This makes one ask why he devotes his life so unstintingly to a dying religion. He regards Hinduism as so meaningful and important that he and Hansar are willing to do all they can to keep it alive for as long as possible.

Bram's father came to Trinidad from India at the age of 18 or 19. Despite the fact that he had a profound influence on the spiritual development of his sons, not one of them actually became a pundit like him. Bram claims that he became aware from an early age that he himself would never be able to live up to being a pundit. So strong was this sense of inadequacy (which undoubtedly still exists) that he soon abandoned all such ideas.

As a child and young man, he was more attracted to Christianity in the form of Presbyterianism, and, although he never actually became a Christian, he claims to know much more about the Bible than the Hindu scriptures—hence his ardent desire to read and understand the Bhagavad-Gita. He attends Mr. Bhattacharya's course in San Fernando and at Débé. Bram reckons that quite a few Hindus, who converted to Christianity, are now drifting back toward Hinduism. He deplores the rift that results from these conversions, because as soon as Christianity has been embraced, Hindus tend to lose their identity as Indians and become creolized. He wishes there were more broad-minded Christian Indians, like George Sammy, who would still have time and sympathy for Hinduism.

Bram regards himself as low-class and financially inferior, despite being a Brahmin. His father's name was not Maharaj but something like Rampersay. When, as a boy, Bram went to the Presbyterian School, the teacher assumed that he must be called Maharaj (great ruler) because his father was a pundit. This name stuck, and was also attached to his brother Hari.

As already noted, Bram has, particularly in middle age, become deeply influenced by his pundit father, who died in 1939. Bram's father used to tell his children morality stories, and he advised them never to have friends, advice which Bram follows assiduously. The fear of being exploited by friends is intense. On the other hand, his father urged him always to do all he can to help other people. He said, "If you can ever be of service, then you must do it." This forms the basis of Bram's character.

Bram's marriage has not been a success. His wife and six children live in San Fernando, while he stays in the family home with his brothers, their wives, and families—a situation that brings to mind the Tulsi home in Naipaul's *A House for Mr Biswas.*[8] He supports his wife and children fully. Nearly all his spare time is spent in and around the Todd Street Mandir, attending Gita and Sanskrit classes and helping Hansar in SSDM.

He spoke about Mrs. Deabi Persad Maharaj (Mistress Deabi) and her brother-in-law, Binie Maharaj, whom we met in the *mandir*. Mistress Deabi has a business on High Street, and Binie Maharaj owns a race-horse that won the Gold Cup. To look at these two you would never imagine that they are wealthy. Their clothes are identical to those of their poorer friends. However, Mr. Binie has a certain air of comfort about him, and he does come to the *mandir* in a huge chauffeur-driven car. Bram seems fond of, and friendly with, Mistress Deabi. He is deeply impressed by her kindness to him and her service to others, especially since he regards her as being in every way his superior.

Mistress Deabi heads a group of Hindu women (Stri Sevak Sabha), who make it their business to see that all Hindu festivals, no matter how small, are observed. Mr. Binie is the financial pillar of Hinduism in San Fernando. He claims to have a good knowledge of Hindi—and so, for sure, does Bram.

Bram also spoke about the social behavior of blacks and Indians. He regards Creole social habits as lax, particularly with regard to sexual behavior, which creates a gulf between the two groups. He is obsessed with protecting his daughter from being seduced.

Colin

It seems that the East Indians in Esperance, where Bram lives, are hoping to build their own temple in the near future. Bram thinks that his brother, Hari, will put up most of the money. Hari is a staunch member of the Sanathan Dharma Maha Sabha (the orthodox Hindu sect in Trinidad), and is, I think, on the Executive. He is, on his own admission, a supporter of the Bhadase Maraj faction.

Bram told us that both he and Hansar are poor Indians, each earning about BWI$60 (£12) a week. Bram is also expecting to contribute something to the upkeep of his brother's family while that brother is out of work. The brother is about 45, an age at which it is difficult to find regular employment. Bram has recently sent to India for two books. One is an annotated edition of the Bhagavad-Gita; the other, a translation, with notes, of the Ramayana.

He still doesn't rule out the possibility of becoming a pundit, though, if he did, he would keep his regular job at Texaco. Bram is hoping one day to take a holiday outside Trinidad, but he will not be going to the beaches of Grenada or Tobago. Instead, he would like to go to Suriname to see East Indian culture; he understands that East Indians have retained their culture there to a high degree. He would like to take a lady friend with his brother and his brother's wife.

Bram has invited us to the wedding of Mr. Ramnath Maharaj's daughter in San Fernando on April 12. I should note here that Bram is familiar with the families of Mr. Ramnath Maharaj and Mistress Deabi—though he regards both as his social superiors. So they are, in a class classification, but they are all Brahmins. Is this significant? Certainly, Bram values the friends he has met through the *mandir* and he would not give them up if a temple were built at Esperance.

Saturday, March 14

Colin: *Tribulations of a Dougla*

I went in the morning to Couva (fig. 3) to look for the man who had run into the back of our car. The taxi driver regaled me with his tales. A *dougla*, he is related on his father's side to a well-known East Indian family. His mother, a Spanish Creole, brought him up in Siparia. When he was young, he used to get a great deal of fun going to the church of *La Divina Pastora* and collecting money thrown under the floorboards by the Indian women. The money was thrown with the first scalp locks of male children on Holy Thursday and Good Friday.

His father owns a small coconut plantation of about 50 acres. He worked there for a while when he left Catholic secondary school. Then he labored on the roads as a line-painter before going to sea, sailing on the run down to British Guiana. When he was on leave at Carnival time one year, he met his future wife, who was a nurse. She became pregnant and married him in 1958. "I never catch she offside!" He then worked as a paraffin salesman, and about two years ago became a driver. His wife still works as a nurse despite having three children (all boys). He likes to think that his name will be continued. They have a servant.

Unlike his wife, he is frequently "offside!" Sometimes he takes a girl down to the beach in his taxi—never letting on that he is married. He usually goes with black girls—of a lower rank than his. But although he regards this as the normal pattern, he did say that his first duty was to see that his home does not suffer, especially financially.

He dislikes East Indians; he says this is because of their rejection of him, as he is illegitimate and half black. His wife, for example, is black, and so are all his friends.

Gillian: *Fashion show, Naparima Bowl*

I went with Ena to a tea party and fashion show given by the Red Cross at the Naparima Bowl, in aid of their proposed new convalescent home. The tea party and show could well have taken place in England, and was—not surprisingly—supported by the middle class upward. Tea was traditionally English with the addition of one or two West Indian cocktail-party savories.

At first, the people there were predominantly white; I suspect that all the whites from Pointe-à-Pierre (Texaco oil compound) must have turned out in force. Then there was a huge group of almost whites, fewer browns, and only a handful of really dark people. As the afternoon went on, more and more Indians appeared, presumably for the fashion show. Of these, about half a dozen were wearing saris. Many of the models in the tiny tots and teenager ranges were East Indian. They make good models, as they are mostly tall and slim, with long black hair. I had to leave before the adult models appeared.

Eileen Appleton (wife of Clayton and a creolized East Indian) came with us, as did Mrs. Winford. Both were impeccably dressed. It is safe to say that nobody who was nonwhite attended this function who could not afford the very latest European-style clothes. Infinitely less well-groomed and chic were the whites.

Jean Mitchell, the compère, brought some life into the show, and managed to put both audience and models at their ease. It turns out that she is Humming Bird herself (the social columnist) of *Trinidad Guardian* fame. The clothes being modeled were lovely—modern and imaginative.

Puja at Timital

On Saturday evening we went with SSDM via La Romaine to Timital, San Francique, to the west of Penal, to perform a *puja* in deeply rural surroundings. We had been led to believe that we were to witness a village *puja*, but I think that it was organized largely by the family under whose house it was held. The people were friendly but much poorer than those of the week before.

One family sticks in my mind. A daughter sat beside me, pointing out her brothers and sisters, who were wearing clean, unpretentious clothes; the *oronhi*-wearing mother, on the other hand, wore an old blouse and patched skirt that had patently been laundered hundreds of times. The girl was fascinated by my short hair; she told me that Indian girls are not allowed to have their hair cut short, as it is regarded as ugly!

The ceremony was brief, and we failed to reach the *hawan* (or sacrificial part) since adequate preparations had not been made. Instead, everyone was invited (again by the pundit, not Hansar) to come up to the altar and perform *aarti*, while *bhajans* (hymns) were sung in Hindi in the background, accompanied by a harmonium, *dholak* (two-sided drum), *tabla* (one-sided drum) and *dhantal* (iron rod and striker).

Here in Trinidad the most popular image is that of Krishna. One by one, people came up and performed *aarti*. Some also moved the plate around above the heads of Hansar, his assistant, and the congregation, asking for God's blessing. It is worth recording that, when the pundit invited people to go forward and perform *aarti,* he also asked that those who do not worship God in this way should worship Him in silence where they sat.

The *puja* came to an end with two sermons preached by Mr. Sookhoo and Mr. Ramsumair from La Romaine. Both addressed the audience in English, though Mr. Ramsumair read and spoke a little Hindi. Mr. Sookhoo talked at length and in an abstract manner. In essence, he was appealing to people to hold on to Hinduism. Hindus here are worried about those who are "blackmailed" into leaving their religion in order to acquire a better job, for example, to become a teacher. Mr. Sookhoo himself is, of course, a rare example of a practicing Hindu who is teaching in a Presbyterian school.

The second sermon had greater popular appeal because it relied less on argument than on a strong narrative. Mr. Ramsumair told the story of Krishna, comparing his birth and youth with that of Jesus Christ. Both were born into a situation of oppression, in which the life of the young child was in danger. Mr. Ramsumair has had little formal education and yet he possessed a marvelous turn of phrase. He held his audience spellbound, and delighted the Hindi speakers by reading an extract from a holy text. The pundit had intended preaching an additional sermon in Hindi, but it was late, and several people were fast asleep by the end of the second sermon, though they were soon woken by a rousing hymn and accompanying music.

During the *puja* I had a good look around me. We were close to cane fields, and away to one side of the house the sky was aglow, from the burning of the cane trash. Men and women tended to sit separately for the *puja*, as they do in the *mandir.* A large number of children were present, many of the little ones fast asleep in their parents' arms or on the ground, covered with a small wrap.

One gray-haired lady, sitting against the kitchen wall, was imperturbably suckling a young baby. Halfway through the *puja* another woman nipped off for a quick smoke by the entrance to the kitchen. Throughout, men and women would stand up, take a breather, and then come back again a few minutes later. Unlike last week, there was no electricity in the house, so the whole scene was illuminated by an oil lamp, suspended from beams supporting the floor above. Every now and then, a family member came and pumped more oil into it.

After the *puja* we got into conversation with some of the worshippers. One man of about 60 told us that he had come over from India at the age of three. As he was so young when he left India, he had no memory of it at all. He didn't even know which part of India he had come from. But he was most talkative, and he told us how much he enjoys smoking ganja.[9] His friends looked askance, warning him against it, but he maintained that it does no harm at all if smoked in moderation.

Then we were taken upstairs into the body of the house for a meal. Before describing the meal, I should say something about the house itself, which had been built recently, entirely from local products. The tall posts on which it stood were tree trunks, the rest wooden planks, unpainted inside, as is the custom here. Perhaps the most interesting feature was the kitchen, a separate, single-story mud construction with a thatched roof—an *ajoupa*. Bram told us that the wattle walls are erected first, and then plastered with fine mud. The final effect was completely smooth and firm. The floor under the main house was made from *gobar* (cow dung, often mixed with mud), carefully prepared, and smoothed repeatedly until the finish resembled concrete.

Before we started eating, a girl came round with a bucket and a bowl of water. In turn, we held our hands over the bucket as she ladled fresh water over them. We rubbed our hands, and a towel was handed round for drying them. Then food was served. We had rice, *dal puri* (lentil roti), salad, pumpkin mayonnaise, and a bean mixture—all very hot but appetizing. When we had finished eating and had drunk our glass of water, the hand-washing ritual was repeated.

Gillian: *Merle Ragoobir*

Merle Ragoobir, one of the SSDM girls, eats meat, but not pork or beef. Like many Indians, she loves to talk about wearing the sari. When I asked whether her Creole friends resented her wearing a sari, which is, after all, Indian rather than Trinidadian dress, she replied that they all admire it. Although she wears the sari only on special occasions, she always wears it when traveling abroad, presumably out of a conscious or even subconscious desire to pass herself off as genuinely Indian. Merle loves her contacts with India: her sister is going to India later in the year, and Merle intends to send one of her finer saris with this sister so that it can be laundered properly!

Merle dislikes the old Indian custom of arranged marriages, and has every intention of choosing a partner herself. But will caste be important? Will she be influenced by her father's wishes? She believes that some form of birth control will have to be introduced in Trinidad. Like Rosalind Ramsamooj and Miss Ramkissoon, she has no desire to have a large family. As Rosalind says, "You can give two or three a chance, but when you have ten or eleven it is a struggle just to keep them clothed and fed, never mind taking an interest in them at school."

Sunday, March 15

Colin: *Todd Street Mandir*

The sandhya puja in the Todd Street *mandir* followed the previous pattern, except that all the children were asked to go up to the images of the gods to perform *aarti*. Most did, though a group of teenage boys refrained. Again I was impressed by the voluntary segregation of the sexes that takes place in the temple. The women and most of the girls (and just a few of the boys) sit at the front and on the left. It is the same at Mr. Bhattacharya's lecture. The core of men is always seated on the right and at the left rear of the *ashram*.

Hansar's address was good. The first time we heard him, he talked about the first two *mantras* of the *puja*, "Let us be saved together, let us be reared together, united and strengthened together too. Let us not be jealous of each other." On this occasion, he talked about worship, and idols as symbols—as means and not as ends.

Visiting the Hercules family, St. Augustine

Later that morning we went with Ena to visit Will and Ena Hercules at St. Augustine, to the east of Port of Spain. I was interested to see how many East Indians there were at the junction of the Southern and Eastern Main Roads. On our journey down the Eastern Main Road, we passed a big mosque at San Juan, the headquarters of the Sanathan Dharma Maha Sabha (SDMS), and, later, Bhadase Maraj's[10] house. Bhadase, it appears, used to have a reputation as a gang leader and thug in the Chaguanas area.

Other guests at our hosts' magnificent new house were Cyril Braithwaite, his wife and stepson, a doctor at the Princess Margaret Hospital in San Fernando. Mr. Braithwaite is an alderman of the borough of San Fernando. He works as a stores controller at Texaco; his wife runs an expensive clothes shop in the arcade. In addition, he also manages a small plantation for a friend. He was offered the rank of alderman to entice him on to the Council. The Borough Council is entirely PNM. Opposition from the DLP and independents has been eradicated. Consequently, a clique within the party can control affairs—sometimes to the embarrassment of other councilors, who if they are not fellow travelers are forced to quit. It seems that many people now hope that the DLP will get some seats at the forthcoming election. Mr. Braithwaite says that, if the opposition knew what he knows about the PNM, they would oust them!

During lunch, an argument developed in the kitchen between Will and Mr. Braithwaite. The former, who is pro-British, was saying that the West Indies has no writer to equal Shakespeare's status. Mr. Braithwaite, however, was championing Sparrow, saying that he is symbolic of the newly independent country and truly Trinidadian. Eventually, conversation got round to color. Will said that, when he was a boy they used to end each school day with "Lord, lighten our darkness, we beseech thee." It had a special social significance for them. Mr. Braithwaite replied that he had attended a Catholic school and had never said "Lighten our darkness." But, as Will retorted, he did pray to a white virgin!

Will, ever the cynic, said he could not foresee a time when color would not play an important part in the social structure of Trinidad. "All dark men choose lighter wives. You did," he said to Mr. Braithwaite. "So did you!" was the spontaneous reply, and both men had to agree. Only Mr. Braithwaite's doctor son said that it was up to people in important positions to change this.

The conversation then moved on to East Indians. Will says that there are only three non-Indian barristers in San Fernando, and one of them (Archbald[11]) is married to an East Indian and is accepted by the East Indian lawyers. Mr. Harry Pooran[12] is obviously regarded as one of the great Indian show-offs. He talks money and income-tax dodges. When we met Harry Pooran (on an earlier occasion), he told us that his father had come to Trinidad in 1903. He has visited his father's younger brother—about 60 miles north of Delhi—and sends him about BWI$60 (£12) per month. He says that the poverty there is unbelievable. Mr. Pooran's father left a wife in India, who was taken by a younger brother.

Monday, March 16

Gillian

When one of Mr. Bhattacharya's pupils asked him what color Christ was, he replied, "My daughter, I do not know what color He was. All I know is that He is now white. He is like a statue that has been whitewashed so many times that people have forgotten its original color."

Tuesday, March 17

Colin: *The Deabi Persad Maharaj family and Mr. Binie*

At 6 p.m. Bram arrived to go with us to Mistress Deabi's house at the corner of Rushworth Street and Prince of Wales Street. Here we were introduced to Dr. Haricharan and his doctor wife Dolly (Mistress Deabi's daughter), and Mistress Deabi's younger daughter, Polly (a teacher). Dr. Haricharan, who is from British Guiana, met his future wife while they were both medical students in India. It appears that university fees in India are very low, which provides an inducement to study there.

The house, painted a vibrant pink, is made of concrete with a brick-like pattern worked into it. On the street side, it is shaded by a broad veranda. The furniture in the main room is comfortable, with an unusual boomerang-shaped pink settee. The wall in this room is covered with religious pictures of Radha and Krishna, Hanuman, Shiva, Rama and Sita that are so ubiquitous here in San Fernando. A basin for washing hands was fixed to the wall separating the dining end of the room from the kitchen.

Bram, Polly, Gillian, and I sat down to eat. None of us used cutlery, though knives and forks were laid for four. The food consisted of rice, curried mutton, salad, potato, mango *anchar* (a peppery preserve), peas, pumpkin, and *dal puri*. Bram was, I think, unhappy about the mutton, and though he took some, I can't remember seeing him eat it. We all drank beer, including Dr. Haricharan, but Mistress Deabi had to press it on Bram, and Gillian said he looked pained when he drank it. Binie Maharaj came in and had supper. He lives next door, and said that his younger brother, Mistress Deabi's husband, was at his house.

Binie Maharaj's father came to Trinidad from Kasiji (Benares, Varanasi) in about 1893. As soon as his indenture was over, he moved to live in Rushworth Street, very close to the present house. In those days, Mr. Binie's parents both worked in the cane fields that stretched up to Rushworth Street. Later, Mr. Binie's father acquired a small coconut property. In 1928, Mr. Binie's older brother opened a store on High Street, which is still run by Mistress Deabi and her husband. Mr. Binie, a tailor by trade, moved into the cinema business in 1938, and since then has made a fortune. He claims he cannot think of a single cinema in Trinidad that is not owned by an Indian.

Mr. Binie is about 56 years old. He is almost bald, rotund, and of dark coloring. (I understand from Miss Rampat that his wife is very light.) Mr. Binie describes himself as retired, though his income is reputed to be about BWI$1,500 (£300) per week—without his racehorses! He gave more than BWI$5,000 (£1000) for the building of the temple in San Fernando. He owns about six racehorses, and at Christmas won the Gold Cup. Mr. Binie reads and speaks Hindi and often goes to Bhagwats to give translations of the readings. He claims to be in the Hindu stage of *dharma* (religious duty) and is patently intent on appearing devout.

A temple seems to have been on the site of the Todd Street *mandir* for the last 13 years. There is also a corrugated iron structure that served as a temple still standing on Prince of Wales Street. The Todd Street temple is run by a registered committee; they also erected the statue of Gandhi. They have leased three lots from the St. Madeleine Sugar Co. at a peppercorn rent for 99 years. The company employs a large number of East Indians, and their support for the temple is felt to be a form of repayment for services rendered. Mr. Binie claims that the new residential area south of Rushworth Street was taken up by people who had moved in from rural areas to ensure an education for their children. The company sold land at about $20,000 BWI per acre, and bought fresh land for cane at approximately one-tenth of this price.

Mr. Binie also frequents the Hindu temple in St. James, Port of Spain. They are very well organized and publish a monthly program of their activities. Simboonath Capildeo,[13] the politician, holds a Gita class there. He is acknowledged to be the best Sanskrit scholar in Trinidad. Neither Bram nor Mr. Binie seems to have any real answer to the competition from Christianity. I have already recorded Bram's pessimism. But both regret the fact that conversion means severance from Hindu and Indian traditions.

While the television was on, Bram was palpably ill at ease. We were watching a mushy story in the *Dr. Kildare* series. Many of the shots were taken on the seashore and involved love scenes, or scenes showing a girl in a bathing costume. I began to tell Bram about Klass's work at Felicity.[14] He says that Felicity (or Amity as it is called in the book) is known to be Indian. It has only one black and he is acculturated. When I asked whether he had ever heard of a *panchayat* (a group of five councilors), he surprised me by saying "yes." It seems that they have now disappeared, but he attended them with his parents when he was a little boy. Leading members of the village were elected to the *panchayat*, and they sat in private to decide on personal issues (such as adultery), and in public when the issues involved the whole community. They aimed at setting a high standard of morality and were courts of justice, outside the law, but binding on Hindus.

Gillian: *Polly Maharaj, teacher*

I spent most of the evening talking to Polly, one of the daughters. She wears western dress most of the time, and a sari on special occasions. In spite of her western appearance, she considers herself to be part eastern and part western—superficially western but mentally and spiritually eastern. Although she has rejected Hinduism (and in fact all religion as a result of her experiences at university in Ireland) she stressed that she is deeply influenced by Hindu thought and philosophy. She cannot speak Hindi and understands only a little. Her sister Dolly, on the other hand, speaks fluent Hindi, which she learnt as a student in India. Polly graduated from Trinity College, Dublin.

She has been back in Trinidad for only a short time and has not yet readjusted to life here. She finds that racial antagonism has increased enormously since the elections of 1956. The majority of people have superficial values, with too much emphasis on the purely physical. Throughout the evening, she kept stressing the word physical, which she was using to convey sexuality, the importance of beauty, especially in women, the accent on smartness, modernity, and at the same time the complete rejection of moral and spiritual matters.

When I asked her about caste, she did not answer directly, but told me that her parents have always impressed upon their children that caste is not important, that it is more important to be a good Hindu. Polly, however, feels that, despite their protestations to the contrary, her parents are very aware of caste and of the fact that they are Brahmin.

Mistress Deabi married at the age of nine. She has ten children, of whom the youngest is a boy of ten. Her daughter Polly teaches in a denominational school in Princes Town, and one of Polly's sisters is on the staff at Naparima Girls' High School in San Fernando. Polly commented on the narrow-minded attitude to teaching prevalent in Trinidad. Only law and medicine, of the professions, are prized. While all visitors to the Deabi Persad Maharaj household are impressed by Polly's doctor sister, they make no comment about the teachers! Polly considers Naipaul[15] very cynical, but she insists that there is absolutely no room in Trinidad for creative writers or artists.

Polly believes that, nowadays, only girls with little education have marriages arranged for them. Polly and her friends would certainly insist on choosing their own husbands. However, she stressed that Indians in many respects still adhere to their old ways. Her father is unquestionably the boss in the family; no one would dream of challenging what he says. Indian women in the street will rarely walk beside their husbands; instead, they walk behind or slightly behind. There is a growing tension among Indians between those with and those without education.

All the Persad Maharaj children, or certainly the older ones, are strictly speaking illegitimate. Polly, who has "illegitimate" stamped on her birth certificate, had great difficulty explaining to a disbelieving Irish landlady that her parents were in fact married "under the bamboo" according to Hindu, but not civil, rites.[16] Before she left for England and Ireland, her father had to sign an affidavit, declaring that he is in fact her father.

She spoke, too, of the problem of teaching rural Indians correct English; if they are to get advancement educationally speaking, they must be able to read and write Standard English. The problem is exacerbated by the fact that there is no tradition of reading in families. It is apparently difficult to persuade pupils why they should learn to read and how to get started.

She thinks that Hinduism has missed the boat in Trinidad in that it failed to appeal during the last 20 to 30 years, which is precisely when it should have been gathering strength.

Polly finds no consolation in Christianity. Her experiences among the Irish Catholics have led her to the opinion that Christianity can be bigoted and narrow-minded. In the same breath she mentioned the Anglican Church in San Fernando. As far as she is concerned, Catholicism and Protestantism have much in common. She feels, too, that it is a great shame that there are so many conflicting ways to God (this is an opinion, which we have heard expressed many times by Indians here), for in the end they all tend to lose sight of God.

Polly's mother makes it her business to observe Hindu festivals in the temple. In particular she organizes prayers at full moon. Unfortunately, she can find no support from the younger generation. This fits in with the general trend—a hard core of middle-aged and elderly people are devout in their religion, yet even their own children have no interest in it.

Wednesday, March 18

Gillian: *Pearl Rampat and Rosalind Ramsamooj*

Pearl Rampat lives in Corinth—on the eastern outskirts of San Fernando, a village in which Indians and blacks live side by side, and yet she can think of no example of a cross-race marriage. She firmly believes that in marriage one must look for a match within one's own race. However, she was by no means as insistent in her antagonism toward Indian-Chinese or Indian-white marriages. When pushed, she agreed that her hostility is largely physical; she doesn't like the hair of the black, nor does she like their dark color.

This question of color applies to Indians too. She regards her own dark complexion as inferior to that of her much fairer sisters. (Her father had fair skin, whereas her mother is darker.) Skin color plays an important part in selecting a potential husband or wife for one's child. Pearl said that some black-Indian marriages take place on the sugar estates.

Pearl's views, particularly with regard to the younger generation and morality, are old-fashioned. She has never in her life been to a public dance. On the other hand, she has no problem with the fact that most Hindu marriages are no longer arranged, though some fathers still look for a husband for their daughters.

Pearl and Rosalind Ramsamooj have received little formal education, and both work mainly from home. Pearl is a seamstress, while Rosalind helps at home as well as working from time to time as a seamstress. Rosalind hopes to go to Canada later this year. But Pearl has no father or brother to arrange a marriage for her, and Hansar is unlikely to impose an arranged marriage on Rosalind. On the other hand, neither of these girls has an opportunity to mix with eligible young men.

Bram does not know the names of the young women in the SSDM group, despite the fact that they all regularly go out together on Saturday evenings and meet in the temple during the week. Although friendly and respectful, he never addresses a word to them unless he has to. Hansar doesn't refer to his daughter or his sons by name, choosing instead "the girl" or "the boy." Bram at his most intimate refers to me as "Mistress Colin," though he normally calls me "Mistress Clarke."

Colin: *Hansar and Bram*

In the evening Hansar started to tell us about us about practicing yoga. He performs a personal *puja* at midnight and frequently manages with virtually no sleep. He started SSDM over ten years ago and now claims to have 14 active groups. Hansar is president of the society, which runs Sunday schools and provides a small library service. He performs a *puja* in Corinth on Sunday mornings before coming to San Fernando. They use a small bamboo and galvanized iron shelter that was erected when Mr. Bhattacharya gave a Gita lecture.

Hansar is a man of immense resources and ability. A poet[17] and songwriter, he also translates. He showed me an attractive card, with a verse inside, which he had made to send to friends at Diwali. Hansar told us that some Brahmins will not accept food from the hands of Hindus other than Brahmins, but that the "take it or leave it" attitude was rapidly becoming prevalent here. Hansar is an excellent speaker, with the ability to put the metaphysical into simple, concrete terms.

According to Bram (on his own), Hansar is an Ahir (grazier) and therefore not of the priestly caste. He has been taking the Sunday service at the San Fernando *mandir* for the last two or three years. Some time ago, a Brahmin priest in San Fernando wrote a letter to other local Brahmins objecting to the fact that the Sunday worship was being conducted by a non-Brahmin. The other Brahmins disagreed with him, retorting that if he objected he should take his own services at a time when the temple was free.

Thursday, March 19

Gillian: *Princes Town and Sigoolan Maharaj*

Bram, Colin, and I drove to Princes Town in the evening to hear a lecture on yoga, which was to have been given by a Creole (Dr. Baldwin George)—much to the delight of our Indian friends. Unfortunately, we discovered, on our arrival, that the talk had been postponed. However, we met Hansar, his daughter Rosalind and Miss Ramkissoon, and we all went for a drink at the home of Sigoolan Maharaj.

It was Sigoolan Maharaj who had invited us to the talk in the first place. He owns a business on the left as you enter Princes Town, immediately after the police station. It comprises a variety of sections (like Lee Chow's emporium in Moruga) including dry goods and a pharmacy, which is sublet. He himself lives in a beautiful old wooden house with delicate fretwork, and has a living room with the usual religious pictures, some of which came from West Germany. His brothers own property behind and around his house.

Mrs. Maharaj was thirteen when she married Sigoolan; she must now be in her forties, though after producing ten children she not surprisingly looks older. As we drank our soft drink, we were shown slides of Sigoolan's visit to India on his world trip. He had taken his father's ashes to be scattered on the waters of the Ganges. The slides were largely of religious—Hindu and Muslim—buildings.

The family showed us their *puja* room; this is a room, frequently at the front of the house, devoted solely to the worship of the Hindu deities. Bram says that when a family cannot afford to set a whole room aside for the gods, they often create a small *puja* "room" in the front garden. In this case, however, the *puja* room was located at the heart of the house. The altar, directly opposite the door, was covered with red cloth, above which there hung a yellow satin canopy. On the floor in front of the altar lay two deerskins for kneeling on; these are kept wrapped in a carpet when the room is not in use.

Around the walls and on the altar were displayed pictures of the deities. In addition, there were several images of the gods on the altar, all of them from India. The central statue was that of Krishna, a *mala* (garland) hung around his neck. The gods portrayed in the pictures and statues were: Hanuman, Shiva, Lakshmi, Radha and Krishna. On the floor there stood a clay incense container, *dias,* and a brass *tarriah.* The family is intensely proud of the peacock feathers hanging on the wall, which are used for fanning the deities and were, like the deities themselves, brought from India. Fresh flowers are put into the room on Sundays, the day on which they perform their *puja*. A *dia* is lit every evening at six o'clock.

There was a tulsi tree[18] in the garden. Apparently, there had formerly been two, but one has been eaten by ants. Every Hindu family, according to Bram, should have a tulsi tree. They also had some faded prayer flags at the front, a relic from their annual *hawan.*

Ramayan satsang at Palmyra

On our way back to San Fernando, we drove past the temple at Palmyra. As a reading in Hindi from the Ramayana was in progress, we decided to stop and see what was going on. The temple itself, erected around 1950 as a result of communal effort, stands somewhat lower than the road at the foot of a steep slope, and is of a typically rural design. Wooden uprights support a vast galvanized roof, and the whole structure is enclosed within a waist-high, mud-covered wattle-and-daub wall. The mud-and-dung floor inside the body of the *siwala* was meticulously tended by one of the village women. Below the roof were hung at ceiling height, crisscross, line upon line of colored paper friezes, all hand-cut.

The temple consists of a large central space with mud floor, in which the readings are held and where all services take place. In one corner, however, there stands a wooden shed, the *mandir* or sacred part of the temple. This contains all the pictures and sculptured images of the gods, and of course the *lingam*. Here *aarti* is performed, and nobody wearing leather footwear may enter. As in all Hindu temples, there were many flowers, though most of them were of plastic.

The pundit and some of his helpers took us into the *mandir* in order to explain what was there and why. The pictures around the altar were of Ganesh, Shiva, and Hanuman, the last clearly being the main god of that temple. Hanuman, a North Indian non-Aryan deity, is much loved by the lower castes. Further pictures showed Narayan and Lakshmi together, Lakshmi on her own, Radha and Krishna, in addition to Rama and Sita.

In a prominent position before the altar rose the *lingam* with the *mahadeo* stone on top. (A picture on the outer wall of the sacred part of the temple actually showed Mahadeo rising from the lingam). Hansar told us that the *mahadeo* stone will grow if constantly fed with water. In the main part of the building, near where the pundit sat, a mud mountain, topped with a yellow flag, had been built. This is done every time there is a reading from the Ramayana, leading up to the Gobardhan puja. Around the mountain there wound a spiral stairway of mud. Normally, a *dia* is placed on top of the mountain.

The officiating pundit sat near the mountain on a yellow cloth. Around him, on pieces of cloth of different colors, sat four men who actually did the reading and chanting for the *satsang* (true company) from Tulsidas's Ramayana.[19] They would read and chant one or two *slokas* (stanzas) in Hindi, and then the pundit added a few comments by way of explanation to those who were listening. Throughout the readings, chantings, and commentary not a word of English was spoken. In all there were 17 people present, including the old pundit. Of these, ten were women—all over the age of 30. Men and women, sitting along the wall, were segregated by sex.

The readings from the Ramayana take one year to complete. This group meets every Thursday evening, and often sits all night listening to the readings and chantings. Their regular weekly service takes place on Monday evenings. They all love hearing the Ramayana, and some of the women were so familiar with the text that they from time to time joined in with the chanting. For them the Ramayana is a guide to the good life. Both Bram and Hansar confessed that they feel happier with the Ramayana than with the Gita, as the former is easier to understand. Even so, Hansar compares the Ramayana with a forest in which it is hard to see the wood for the trees!

The *rehal* (wooden bookstand) on which the Ramayana was placed, had been cut from a single piece of wood, and was elaborately carved. The wood had been split down the center, and the two pieces cannot be separated. They interlock and open up to form a book rest, which stands on the floor; the holy text, when opened, is thus easily visible to the reader, who sits cross-legged on the ground in front of it.

The pundit, a man of about 65, was visiting the temple at Palmyra for the first time. He spoke fluent English, though Hindi was his native tongue. He seemed delighted that we were taking an interest in Hinduism, and it was he who explained to me at length the identity and significance of the various deities. He wore a *dhoti* and *kurta*, a saffron scarf, and the sacred thread around his neck. On his brow were three vertical stripes in red and white, and just in front of his ears he had round white marks. He snatched a quick cigarette as often as possible!

Colin

The temple is associated with the Sanathan Dharma Maha Sabha sect—the orthodox Hindu movement in Trinidad. We were introduced to most of the men. Five out of the seven had a surname associated with the names Singh (implying membership of the Chattri caste) or Maharaj (Brahmin). One of the Brahmins, Ragoonanan, who is a pundit in Palmyra and Mount Stewart, said he was related to Seukeran[20] and Rudranath Capildeo on his wife's side. He has a son working in London and studying economics part-time.

A Mr. Singh, who works at Texaco and knows Bram, has a son studying medicine in Georgia. One of the other Brahmins is the grandfather of the Sigoolan Maharaj children in Princes Town. He is a vegetarian and does yoga. I discussed *gotra* (clan) with Hansar and these two men. It was agreed that only Brahmins know their *gotra*, and only pundits, with any real certainty.

Hansar's home, Corinth

We later stopped at Hansar's home at Corinth. The house is big enough for two families (originally three). As is the tradition here, it has been built on stilts and constructed so that extra rooms can be added at a later date. Subdivided inside into a series of cubicles open to the roof, the main room was tidy, and contained three or four easy chairs, a refrigerator, and pictures of Hindu deities as well as of teams representing Oriental Cricket Club. We sent out for potato *roti* and chicken and had soft drinks. Although Hansar and Bram eat meat (but not pork or beef), they remain strictly vegetarian on days when they attend religious services. Bram drinks alcohol very occasionally—but his youngest brother (the black sheep of the family) used to be a regular drinker.

Saturday, March 21

Gillian: *At home with the Bhattacharyas*

In the evening Bram went with us to Tagore College at Craignish, immediately to the east of Princes Town to have a meal with Mr. Bhattacharya. The college commands an unbroken view of the rolling cane land back to Naparima Hill.

The building itself is square, with the schoolrooms of the college on the ground floor and Mr. Bhattacharya's generous flat above. The main room, a beautifully furnished study-cum-living room, is big, with a high ceiling. It is the most elegant and imaginatively decorated room we have seen since our arrival in Trinidad. Around the walls there were tall bookcases bulging with books. Above them, at picture-rail level, were displayed Mr. Bhattacharya's own paintings. Apparently, he has been painting only for four or five years, since having a nervous breakdown.

Mr. Bhattacharya's main loves are literature, art, and music. The Indian music played in the background later in the evening was delightful. (Mr. Bhattacharya's brother is a leading composer in India). Instead of mass-produced religious pictures, he had some finely carved images of Hindu deities, worked in a clear white stone, and some black reliefs. All very refined and restrained. Unusually, Mr. Bhattacharya wore trousers with an open-necked shirt, and he seemed more subdued than he is in the *mandir*.

Within a few minutes of our arrival he confessed that he suffers from egocentricity, an admission with which nobody could disagree. But I have never come across a less selfish egocentric. He knows that he is intelligent and forthright. He offends people wherever he goes, and yet he has integrity and can face up to opposition when he is convinced he is right. Moreover, he is doing some great work here in Trinidad with regard to the propagation of a new, revised Hinduism. Or rather, he is examining traditional Hinduism closely and suggesting where it has gone wrong.

Mrs. Bhattacharya is an almost silent, traditional Indian wife. Though shy at first, she was more at ease by the end of the evening. On her brow she had a red dot; in the parting of her hair a *sindoor* (vermillion) line. Unlike her husband, she speaks English hesitantly.

We also met a Dutch couple, Dr. and Mrs. Dan van Eendenburg, from Shell at Point Fortin. Dan, who is very knowledgeable about Hindu philosophy and scriptures, gives Gita classes in Point Fortin. He stressed the importance of caste among Hindus, in spite of the fact that it would appear to be of little significance today. He claims that it is difficult to gauge its importance, as little is ever spoken about it. Mr. Bhattacharya, too, stated bluntly that he considers caste very important among the Indians, particularly among the Brahmins, who are determined to preserve the status quo.

Mr. Bhattacharya underlined the failure of the Trinidad pundits to give any spiritual lead. A pundit is supposed to teach people, but all too often, pundits, who themselves are ignorant, have been determined to keep to themselves what little information they possess. The result is that the people are spiritually starved and can only go through the motions when performing rituals, without understanding their significance.

He commented on the difference between Hindu practice in Trinidad and in India. The ceremonies in the temples here and the forms of private prayer tend to be low caste. Hanuman, as we had already seen at Palmyra, is a much-loved deity among the Trinidad Indians. In India, Hanuman is the deity favored by the lower castes. Hinduism is a much more personal religion in India, and is free of many of the embellishments found here. Mr. Bhattacharya commented on the habit of greeting fellow worshippers as one enters the temple in Trinidad; in India this does not happen. The most common greeting encountered here is not *Namaste* but *Ram Ram*. The greeting *Ram Ram*, with its response *Sita Ram*, is more commonly used by the lower castes.

Dan stressed the difficulty of getting to the truth about Indians and Hinduism. He urged us not to be misled by the number of high-caste names; in his opinion, the name Maharaj is frequently assumed by those who want to be thought of as belonging to a higher caste than they are. He believes that the majority of Hindus in Trinidad are of low caste. Money is the only way of getting around the problem of caste. If a man of lower caste wants to marry a Brahmin girl, for example, he will certainly be rejected unless he has a lot of money. Color too is significant; he will stand a better chance if his skin is fair and his features Aryan. Dan considers that the Indians, particularly the Hindus, are far too divided among themselves.

Mrs. Vilma Dubé,[21] the wife of a Princes Town doctor, was also present. Although Christian, Vilma is interested in Hinduism and Hindu philosophy. According to her, Mr. Bhattacharya is far from being a denouncer of Christianity, as I had thought; he is impressed by much of it. He does, however, feel that like most religions, Hinduism included, the essence of Christianity is obscured by a mass of dogma and ritual.

Vilma studied Indian History and Philosophy at the University of Lucknow. In dress, appearance, and faith, she is (like Polly Maharaj) completely western, but this has brought with it no rejection of, or scorn for, Indian culture. She is the antithesis of Grace Kangaloo and Eileen Appleton; in comparison with them she seems very Indian (eastern), yet in comparison with Mrs. Bhattacharya she seems utterly western!

Vilma has several sisters, two of whom teach in San Fernando—one at Naparima Girls' High School. Vilma likes classical music and greatly admires a Lebanese-American poet, Khalil Gibran, who illustrates his own work. All three (Mr. Bhattacharya, Dan, and Vilma) admire the writing of V. S. Naipaul,[22] and in particular *A House for Mr Biswas.*

I can scarcely begin to describe the food we ate. It was displayed on a long refectory table, at which we sat. Cutlery was provided for everyone, though Mr. Bhattacharya and Bram soon abandoned theirs in favor of their fingers. Herring and chicken were served. In addition, there were some tiny balls of what looked like finely minced meat. In fact this was a vegetable dish made from the flower of the banana. We had plates of rice in a vegetable curry, two different kinds of light, puffy *roti*, vegetables, and something made from milk that looked like tiny fried potatoes. In short, there was a banquet of delicious, subtly flavored dishes, to which we could add any number of sweet and spicy chutneys and sauces.

As if this wasn't enough, there were three different desserts. The first was a fruit purée with what seemed to be chunks of apple. This was followed by white balls made from milk boiled over and over again with sugar, and green pawpaw fried in *ghee.* I thought it worth mentioning exactly what we ate, as I have never before had such a feast. Mrs. Bhattacharya, like Mistress Deabi, did not eat with us. She remained in the background at first, and we were served by her daughter and Vilma.

Dan stressed that we should not take anything we saw or experienced in the Bhattacharya home as typical of Trinidad Indian life. Mr. Bhattacharya and his family are high class, high-caste Indians. Their food and way of life are completely different from that found in even the most sophisticated Trinidad Indian home.

Colin

Mr. Bhattacharya is left wing, and an admirer of the British Labour Party, who finds Britain truly democratic. However, he was a saboteur in India as a young man, and sat for his MA degree under an assumed name. His father was a yogi and a *guru* (spiritual teacher). Mr. Bhattacharya worked in British Guiana before coming to Trinidad, but suffered a nervous breakdown and took up painting. I particularly liked a watercolor of his, in blue, mauve, and pink, of the swamplands at sunset. On his way back to India via Trinidad, he had lectured in San Fernando and was invited to establish a college—Tagore College at Craignish. His backers disagreed among themselves within three months of his arrival, and he now has hardly any funds. Gopaul[23] (Marabella) finances the school as (I think) a profit-making concern.

Mr. Bhattacharya spoke out strongly against caste differences here. He has a great love of literature, is an admirer of Naipaul, and sympathized with *The Middle Passage*. He likes *A House for Mr Biswas*, but finds it very Russian and too gloomy for the Trinidad spirit. Mr. Bhattacharya is an interesting, intelligent, rather isolated, egotistical yet humble man, who gives himself wholeheartedly to the cause of modernizing the Hindu community.

Monday, March 23

Colin: *More on Mr. Bhattacharya*

We had a conversation with Hansar, who mentioned the antagonism toward Mr. Bhattacharya. At Débé, one of the pundits was annoyed because he could not personally attract an audience like Mr. Bhattacharya. The pundit had the temporary shelter at the rear of the Seunarine Temple pulled down, but now another group claims that they will rebuild it to house Mr. Bhattacharya's Gita class.

Gillian: *Rosalind Ramsamooj and Pearl Rampat*

I had a conversation after the Gita class with Rosalind Ramsamooj and Pearl Rampat. They too spoke of the antagonism of pundits in Trinidad toward Mr. Bhattacharya. But he holds a position of authority that Hansar cannot enjoy, since Mr. Bhattacharya is well educated and a Brahmin.

They also mentioned the antagonism of pundits toward SSDM and Hansar. The reasons for this are that Hansar is not a Brahmin; he does not accept fees; he goes among the people; he helps them to understand the scriptures themselves by translating from Hindi to English, explaining the theology and the meanings of prayers and ceremonies. What's more, Hansar preaches that caste is not significant. The important thing is to be a good Hindu. Why should only Brahmins become pundits? According to Bram, Hansar is an Ahir. There is a saying that "you can no more get butter out of sand than good from an Ahir." SSDM is undermining the authority of the pundits and jeopardizing their hitherto unchallenged position as Brahmins.

The two women feel strongly that Trinidad pundits have failed in their duty during the past 30 years. They only have themselves to blame for the position they are in. Much depends on one's definition of a pundit. Hansar regards as pundit anyone who seeks divine knowledge, and who makes it his duty to pass this knowledge on to others. This is not, of course, the traditional definition of a pundit. Trinidad pundits, according to Rosalind and Pearl, have in the past hoarded what knowledge they possessed. Rosalind considers us lucky to have come across the SSDM group, because all the group members are willing to discuss Hinduism and its problems. She claims that we would find no such cooperation among the pundits. I hope this isn't true!

They also commented on politics, observing that the present Peoples National Movement (PNM) government is unfair to the East Indians. They claim that the government is not always honest—for example, there is the question of all the money raised for hurricane relief in Tobago. It is widely believed that Tobago has received only a small part of this fund.

The attitude of creolized East Indians toward the rest of the Indian community is little different from that of Creoles; when speaking of "the Indians," they dissociate themselves completely from them. Every time a Hindu wedding is mentioned, the stock Creole comment is, "you get no meat and drink there."

Partying is popular and important to Creoles in Trinidad and, according to Rosalind, is at its very worst in the week before Carnival. She and Pearl have little time for the "fête like fire" mentality, so prevalent here. Rosalind said, and Pearl agreed, that it is no fun to live in a society such as this when you don't hold the same views. Creoles and Indians diverge on religious and moral grounds. Hinduism is looked down on as a religion of poor illiterate idol-worshippers, and their religious views are not respected by society. Hinduism demands, particularly of women, a strict code of moral behavior. Many of its demands are, in the eyes of modern, progressive, western society, outdated and unnecessary.

Many Indians feel that Christians (particularly Anglicans) in San Fernando go to church to be respectable and fashionable. Such a high standard of dress is set that 80 percent of the population of

San Fernando must be excluded for lack of money. Consequently, the Anglican Church is dominated by the upper-middle class. With Hindus, the situation is reversed. Anyone, no matter how rich or poor, can go along in any clothes to worship in the temple. The result is that poorer people go along, and the rich find it more respectable to stay at home, though there are exceptions such as Mr. Sigoolan Maharaj, Mr. Binie, and Mistress Deabi.

Friday, March 27

Colin: *Siparu Mai*

In the evening of Holy Thursday (March 26) Gillian and I drove to Siparia with Bram. We were going to see the Hindu worship of La Divina Pastora, a Roman Catholic deity enshrined in the new parish church and known to the Hindus as Siparu Mai. During the drive, Bram inveighed against intermarriage with blacks—all Indians feel strongly about this. They fear that their daughters will be seduced. Bram says that he carefully questions his daughters when they go to a party, and makes sure to find out which boys will take them and meet them. He agrees strongly that Ramnath Maharaj is right to keep his daughter at home, although she has had secretarial training.

The conversation moved on to caste. All Hindus know their caste[24] and one another's. People will tell you what their caste is. The Arya Samaj[25] (especially the leaders) and some Christian Indians take caste into consideration when marrying. Bram sees caste as inevitable for Hindus—it is enshrined in the text of the Gita. But this does not permit ostracism of the low castes—all have a duty to perform. Bram would have no objection to a child of his marrying a good Hindu (of any caste), or a good, nonfanatical Muslim or Christian.

Beforehand, Bram would check the background of the parents and would advise on—if not arrange—a child's wedding. He and his younger brother (Hari) would have arranged a wedding for their youngest brother if he had not beaten them to it. Hari is the head of the household, not because he is the oldest, but because he is the wealthiest. Not all Hindus have the same pundits; Chamars often have their own. Gosain is the highest caste, higher even than Maharaj. The *saddhu* at the temple is a Brahmin, but the previous one was not. There is still a *saddhu* at the old temple—an *ojha* or *obeah* man called Ramsamooj.[26]

The Stri Sevak Sabha (Women's Association) at the temple has Mistress Deabi as President and another Mrs. Maharaj as secretary.

The entire group that celebrated the new moon recently in the temple seemed to be Brahmin (17 women and three men).

Bram's father (Rampersay) had a Nau (member of the barber caste) who helped him as a pundit. The barber ran errands for him, and cut his hair, fingernails, and toenails, and those of his children. This was a kind of *jajmani* system.[27]

There are many poor people living in the shantytown on the San Fernando dump (Kakatwey). One woman from there, who sells in the market, comes to the *mandir*, but not on Sundays. Among the poor there is still some animal sacrifice, especially of goats.

When we got to Siparia, the church was full of Catholic communicants, and we became aware of the distinction between Creoles inside the church and Indians in the neighboring streets. We waited until the service was over, expecting the Indian crowd to storm the church. In fact, the gates were closed and the porch sealed for the night. We spoke to a Christian East Indian who said that attendance at the festival had fallen off, but that many Hindus would come to Siparia in the morning to pay homage to the saint.

Before returning to the car we took one last turn. At the back of the church, and separated from it by a dusty playground, we found the old schoolroom. We climbed the flight of steps and to our surprise saw that La Divina Pastora was enthroned on a platform at the far end of the room. The platform was about four feet high and extended across the end of the room. It was surrounded by a high barricade of ferns, which was broken immediately in front of the statue. This gap was framed by an arch about seven feet high, also decorated with greenery.

The schoolroom was full of beggars, mostly old or crippled East Indians, though there were a few blacks. They sat in a long line down the room, and it was clear that many had come early to ensure a good pitch for the almsgiving that accompanies the pilgrimage. The sight was pitiable.

We spoke to two of the organizers. They were both Catholics: one black and the other East Indian. Their task was to ensure order, guard the statue, collect the offerings, and help pilgrims on the Friday morning. They said that as much as BWI$2,000 (£400) was collected in offerings made by devotees of Siparu Mai. Bram had become more and more interested in the festival during the course of the evening, so he and I resolved to return on Good Friday morning.

We set out next day at eight o'clock, leaving Gillian to go with Ena to St. Paul's Church in San Fernando. We were afraid that we might have been too late to witness the celebrations, but when we reached the school at a quarter to nine, it was clear they had only just begun. Several hundred people were already in the yard. We were first attracted to a group of about 50 people surrounding two male dancers dressed as females.

One dancer was quite stout. He wore an *oronhi*, shirt, and long skirt. His cheeks were rouged and he had prominent gold fillings in his teeth. The other dancer was slim and rather effeminate, wearing a false bust and anklets. Unlike his companion, his *oronhi* was topped by a crown. His cheeks were rouged and, like his companion, he was wearing lipstick. He wore a blue bodice and a long, flowing pink skirt. While he danced quite elaborately and with hips swaying sexily to the music, the other man would take up children and dance with them. Their mother or grandmother put some money in a handkerchief spread as a receptacle on the ground. The man danced on the *oronhi*, holding the child and singing to the music, which was provided by a harmonium and drums. At the end of the dance, the man bent down, pressed his thumb into the dust, and made a mark on the forehead of both the baby and its mother. In all the instances that I can remember, a baby boy was involved.

We then went into the school by the middle door. Crowds of people were filing through, making gifts of money to Siparu Mai (collected by her custodian, an elderly black lady) and offering sweet oil and candles. We later went to the side of the platform and saw kerosene tins full of palm oil and a huge pile of candles. None of the four or five people accepting these offerings was East Indian. Money was usually handed to the woman, but some coins were thrown to Siparu Mai from the body of the schoolroom. Only half the oil was poured out for Siparu Mai. The remainder, regarded as holy oil, was returned to the donor for his or her medicinal use.

The devotees included people of all ages and both sexes. Almost all were East Indian, and the majority, according to Bram, were Hindu.[28] Bram did, however, recognize one or two non-Hindu East Indians; one family was Catholic and the other Muslim. After praying and making their offerings to the saint, the devotees moved slowly down the line of beggars (some of whom had come from the sidewalks of San Fernando) giving two cents and a handful of rice to each person. Some of the mendicants wore *dhoti* and *kurta*, and one or two had pink turbans and the *janeo*. Possibly one or two were very old pundits. The scene made a strong contrast with the schoolroom's patriotic, but dated, pictures of Queen Elizabeth and the Duke of Edinburgh, and a map of the electoral boundaries of Trinidad.

When we went outside to look for the Nau, we found them in the far corner of the schoolyard. Most were giving male children a "short back and sides." These boys were all little more than babies, one being suckled by his mother immediately after his trim. One small boy, with hair down to his shoulders (indicating that this was to be his first haircut) was completely shaved. He was carried off grizzling and peering out from under a sun hat that had been placed on his bald head. His mother took the hair and carried it to the end of the schoolroom where Siparu Mai stood. She laid it in the dust under the building and covered it up. Then little Creole boys waited until her back was turned and scuffed the hair up. But to their dismay there was no sign of the silver coin usually offered by parents. Again, only boy children were shaved, and I do not recollect that they had brothers or sisters—though they might have been left elsewhere.

Finally we made our way around to the front of the school, past another Nau trimming a man's hair (not all are professional barbers) and into the school. In there we met a couple from Claxton Bay. They were Hindus and devotees of Siparu Mai, whom they clearly regarded as part of the Hindu pantheon. They said that people came to pray to Siparu Mai to help them out of trouble, to give them health, and especially to give them children. Children who were the gift of Siparu Mai were shaved outside the church and offered to the dancers in thanksgiving. At the same time, people also went to the dancers to secure their future health and happiness, and we saw an elderly lady, laden with silver anklets and bangles, and a young crippled boy, both requesting the help of the two dancers. In these cases, the man danced on the woman's *oronhi* and then pressed her shoulders, her knees, and feet—only the latter part was performed for the boy.

The couple from Claxton Bay obviously believe in the divinity of Siparu Mai. About 20 years ago, the woman's mother had been very ill. A friend of hers had come to the festival and had prayed secretly for her. She promised that if she did recover, her friend would have a small silver replica of the saint made, and that she would offer it to Siparu Mai on the next Good Friday. The woman did recover, and when her friend told her about the promise, she decided to fulfill it. A year later she prepared to go to Siparia and to take the statue with her. A day or so before her departure she again fell ill. But she was determined to go to fulfill the promise made on her behalf. She made the journey, and when she returned home, she became well again, and lived for many years afterward.

Bram was fascinated by this festival, and I found his comments valuable. As an orthodox Hindu, he thinks that the devotees have been misled. He blames the Catholic priests and their greed for money. He is surprised that Hindus should have worshipped a saint in a Catholic church during the nineteenth century, when Christians throughout the island were attacking them for their paganism. Finally, he was disappointed that so many people (hundreds if not thousands) should turn out to worship Siparu Mai, while they do not enter a Hindu temple in such numbers.

Sunday, March 29

Gillian: *The Nandlal Hindu wedding*

We went with Bram to attend the Orthodox Hindu wedding in San Fernando of Miss Nandlal to a young man, whose family had adopted a European surname—Foster. Hindu marriage is not normally performed in the temple but at the home of the bride, where a *maro*, or marriage booth, is constructed for the occasion.

In this instance the *maro* had been built over the stony yard at the side of the house. Its framework was of tall bamboo stems, covered by a galvanized roof. Folding wooden chairs were arranged in haphazard rows around a central wooden platform, about six inches high and about seven feet square. Hansar expressed the hope that their marriage would not turn out to be as shaky as the platform on which it took place! In the center of the *maro* stood a tall bamboo pole. This central pole should pass right up through the roof, and the foliage on top should be left on, so that it flutters in the breeze and attracts the attention of God throughout the ceremony. In this case, although it didn't penetrate the galvanized roof, the foliage had nevertheless been left on.

At the foot of the central bamboo, a *vedi* had been constructed; this should always be made of mud and cow dung and decorated with colored rice. In this case, a pattern of colored squares had been worked over a white background. Around the central pole and *vedi* were the *haris* (a piece of wood dyed yellow in turmeric water with five notches representing a ploughshare), sacred fire, a grindstone, and a mortar and pestle. The central bamboo is usually sheathed in banana stems. Here, however, it was wrapped in red paper with thin strips of blue and pink paper wound diagonally around. A banana sucker was planted at the base of the pole. The *maro* was hung with streamers made from colored paper friezes depicting one of the milk-maids associated with Krishna.

Milap

By the time we arrived, the groom's *barat* (motorcade) was parked a short distance away from the home of the bride, and the drummers were beating a steady fast rhythm. Before the groom can enter the bride's house there is a *milap* (meeting-up ceremony), in which the groom's father, followed by the groom's kinsmen and male friends, is formally welcomed by the bride's father. In this case the bride's father, dressed in a white shirt and *dhoti* and carrying a *lotah*, emerged from his gate and, accompanied by his pundit, walked slowly down the hill toward the groom's father and his party. As the two groups drew closer together, the drums beat louder and faster, adding drama and tension to the meeting.

After welcoming the groom's father, who was dressed in a white *kurta* and black trousers with a wide-brimmed black hat, the bride's father led him and his followers in through the gate, where they sat and waited for the groom. Actually, the role of father to the groom was performed by an uncle because the groom's father was dead. From now on, that uncle will continue to be regarded as father-in-law to the bride and will be treated as such by her. He in return will treat the bride as his own daughter. Since the bride had no younger brother to assist in the ceremonies, she had a substitute brother for the purposes of the wedding. This was the young son of a close friend of the family. This boy will now stand in relation to the bride as a brother.

Oddly enough, there was only one pundit rather than the usual two. He was wearing a white *kurta* and *dhoti* with a pink scarf and white cap. Likewise, there was one set of drummers rather than two rival groups. The four drummers were all Indian and probably came from Débé or Penal. They were in no way dressed up for the occasion, though they certainly put all they had into their drumming. The drums were slung around their necks, and a large one was covered with cotton material. The drummers used two sticks or their bare hands, though the man with the bass drum used one hand and a stick (Plate 3).

Parchan

Once the groom's father and party were inside the yard, the car containing the groom and his *siballa* (companion or best man) drew up just outside the gate and waited. The *siballa* is usually the groom's younger brother who acts as a best man-cum-pageboy. After a while,

Plate 3 Drumming during the *milap* (meeting of the fathers of the bride and groom), Hindu wedding, San Fernando

the bride's mother and some of her female friends approached the car, bearing a plate of burning camphor and a *lotah* containing water (Plate 4). The *lotah* of water was passed over the groom's head, and he was offered a small gift of coins in order to tempt him from the car. The women also threw small balls of flour over the car and into the crowd of spectators. Finally, the groom and his *siballa* left the car and entered the yard.

Although the groom wore traditional Hindu wedding costume, underneath it he had on a pair of smart fawn trousers, a white shirt, and new blue nylon socks. All of this was covered with a heavy pink, long-sleeved satin robe (*jora-jama*) reaching down to the ground. On each wrist he wore a heavy silver bangle, and on his feet an ornate, sturdy, round-toed pair of white shoes. (Hansar said that the shoes are traditionally long with curled, pointed fronts).The yoke of his robe was smocked, with tiny gold beads worked in.

The nails on his hands and feet had been painted with pink varnish. On his head the groom wore an elaborate crown, set on a pink satin turban. The crown was made, or is at least traditionally supposed to be made, on a framework of wire and bamboo covered with gold paper. From this gleaming crown there hung colored beads and colored tinsel mirrors. Artificial flowers hung down in ribbons either

Plate 4 Tempting the groom and his *siballa* (companion or best-man) out of the wedding car during *parchan* (ritual greeting of groom at the bride's home), Hindu wedding, San Fernando. Note the wearing of the *oronhi* (veil) by the women

from the crown or the turban, around his shoulders and down his back. These were in green and red; the crown is called the *maur*.

Over his face the groom had a veil. In fact, it was like a mask, from which more strings of paper flowers hung. This time the flowers were white and red. The veil, tied around the groom's head by white ribbon, was taken from him and put over the bride's face when she first entered the *maro*. The *siballa*, sporting a pink satin turban, also wore a long, pink satin gown, though his lacked the groom's ornamentation. In fact, the *siballa* played a minimal role in the proceedings, preferring to spend most of the time on his mother's knee.

Dwar Puja

This ceremony took place on a spot just inside the gateway to the yard. In this short *puja*, conducted entirely by the pundit, the groom is welcomed first by the bride's brother, and then by her father. The significant moments were the gift of money to the groom and the ceremonial washing of the groom's right toe by both brother and father of the bride. The ritual was performed first by the bride's brother, then, having been presented with gifts of a *dhoti*, a *lotah* and *tarriah*, shirts, rice, dried coconut, and sweetmeats, he left the marked out altar and his place was taken by the bride's father, who went through an identical ritual but received no gifts.

No altar had been built up, though one was marked out on the sandy ground with leaves and petals. As usual, the sacred fire burned throughout the *puja*. In the rectangle that signified the *vedi* there stood an earthenware jar, decorated with cow dung and mud, and a *lotah* filled with water, which was sprinkled around using a mango leaf. Around the neck of the jar had been wound a band of cloth similar to the one tied around the wrist of the bride before the day of the wedding.

A ring, formed from the central fiber of the mango leaf, was put on a finger of all the participants (groom, bride's brother, and bride's father) before the *puja* (performed entirely in Sanskrit) got under way. As neither the groom, nor the bride's brother, or father understood a word of what was happening, and what was expected of them at each stage, they were prompted throughout by the pundit, who gave them clear instructions in English. The sticks used on the fire, onto which *ghee* (clarified butter) was thrown, were of pitch pine. During the dwar puja the bride's brother and father, each pressed saffron onto the groom's brow. The bride's brother was wearing a shirt and *dhoti*, which made him feel very self-conscious in front of his friends. Later that evening we saw him looking much more comfortable in jeans.

At this point during the ceremony, we had a pause, during which the groom drew back from the *vedi* and sat on a chair near his "father," until the bride entered and had a small ritual performed for her. During this pause, we had a chat with the pundit, who tried to explain the various parts of the ceremony to us. We asked whether the form of the ceremony is standard for all castes. He replied that some special prayers are said for Brahmins. He did not, however, disclose whether this was a Brahmin wedding. It wasn't.

After the bride had been led into the *maro* she sat down on a chair, facing the *vedi*, with her mother on another chair directly behind her, facing the same way. Thus the mother's legs came round those of the girl, symbolizing (I presume) the birth position. Hansar said that long ago the bride used to touch her mother's breast as a sign of the last moment of dependence on the mother and her family. Once the mother and daughter had settled into position, the wife of the Nau symbolically went through the motions of cutting the nails of mother and daughter; then she painted red dye on their feet. The bride was barefoot most of the time.

Dal Puja

The groom's "father" then entered the *maro* carrying a suitcase full of clothes and jewellery for the bride. Traditionally she is given a complete going-away outfit by her brother-in-law. The gifts were handed to the bride by the groom's elder brother, who, once he had finished, hung a pink woolen garland (like a fancy skein of wool) around the bride's neck. This is supposed to be the first and last time the groom's elder brother has physical contact with his younger brother's wife. The whole incident seemed to be regarded as something of a joke, particularly by the groom's friends. The bride was then led back into the house by her mother and the Nau's wife.

This is where the central marriage ceremony begins. Now the groom came into the *maro* and sat down on a low bench to the left of the pundit. The father of the bride offered him a *lotah* of water to drink, and once more washed the groom's big toe. The *pirah* (low bench), on which the groom sat, is supposed to be constructed without a break or join from one piece of wood. That was not true of this particular bench.

Kanya Dan

At this point the bride was led in again by her mother, father and the Nau's wife. The bride's father sat down to the right of the pundit, with his daughter on his left thigh. This act symbolizes the daughter coming from his body. The bride's little sister, who until now had shown no interest in the proceedings, became jealous of her sister and started to bite her father. She was dragged away from the *maro* by other members of the family. The bride's mother had changed her dress since her last appearance in the *maro*. Instead of floral nylon, she was now wearing a yellow cotton dress with matching *oronhi*. A *tika* was placed on the brow of both bride and groom by the bride's father.

Gupta Dan

The Nau, who had been in the *maro* most of the time and had helped the bride's father in the exchange of money at the dwar puja, handed a ball of dough, containing coins and a hibiscus flower, to the bride's father, who, in turn, handed it to the groom. The bride then held the ball in her outstretched hand, and her brother poured water over it from the *lotah* while the pundit recited in Sanskrit. Sometimes the dough is shaped in the form of a vagina, and the groom in extracting the flower and coins symbolically breaks the hymen. Unfortunately, we couldn't see whether this was the case here. The dough ball was then handed to the groom's father, who fished out all the money.

Pau Puja

The bride stepped around the altar and sat on the groom's right hand, both of them on the long low bench described earlier. The bride's father offered gifts to the pair. Apparently, it is customary to give a present of a *lotah* and *tarriah* on the occasion of a wedding, and some couples receive as many as 30 of each. They can rarely make use of so many, and will probably, in their turn, hand them on to other brides and grooms. All those who had offered a gift to the young couple entered the *maro* and symbolically washed the feet of the bride and groom. This short ceremony is called neuta. The presents should be recorded in the family invitation book, but I don't think they were. All presents other than the *lotah* and *tarriah* were left still wrapped inside the house on the bride's bed. From the outside at least they looked no different from those of any western bride.

Hawan

During the fire ritual, offerings were made of ghee, flower petals and water, and parched rice. The parched rice should always be brought from the homes of both bride and groom, then ceremoniously mixed by the *siballa*. This parched rice is then called *lawa*.

Exchange of vows

Before the bride and groom exchanged vows, the pundit vainly called for silence, and read, to those who remained to witness this part of the ceremony, the marriage vows of the Trinidad Pundits' Council translated from Hindi into English. This was the only part of the ceremony to be translated into English (apart from instructions given throughout to the bride and groom); the rest was in Sanskrit, and incomprehensible to us and most of the witnesses. But few were interested in what was going on. During the translation, the pundit had to turn and ask the bride and groom to remind him of their names.

Saptapadi

The groom's sash was knotted to the bride's clothes. Thus joined, they circled the sacred fire seven times (*saptapadi*) in a clockwise direction, with the bride leading for the first four laps and the groom for the final three. At the end of each complete circle, the bride's "brother" poured a small amount of *lawa* onto a special wooden trident held by both bride and groom. The first four times the groom stood behind the bride, clasping her hands as she poured, then the process was reversed. Each circumambulation represented one witness to the exchange of vows; for example, the sun, moon, planets,

God, and the people were all called upon to bear witness. Once this was completed, the bride sat down on the left of the groom, as his wife.

Sindoor Dan

The bride and groom then swiveled round on the bench to face each other. The Nau's wife entered the *maro* and covered the couple with a clean white sheet. Under cover of the sheet the groom lifted his wife's veil and rubbed *sindoor* (vermillion powder) into the center parting of her hair. Traditionally, this was the moment when the groom lifted the veil and saw his wife's face for the first time.

This moment during the ceremony didn't turn out to be as private as it should have been. The groom had some difficulty in getting the *sindoor* onto his bride's parting and had to be helped by the Nau's wife. Then a number of married women entered the *maro*, and in turn walked up to the bride and greeted her. This is their way of admitting her into the ranks of married women. From now on, according to Hansar, she should leave her maiden friends behind, though this will, of course, not happen literally. Finally, the couple, their garments still tied together, moved off into the house, their hands full of rice.

Kohabar

We failed to witness this part of the ceremony as it took place in a small room inside the house. But we did see the remains afterwards as the floor was scattered with rice. Clearly, the groom's crown was removed from his head, and the knot joining the couple was undone. He emerged a short while later looking tired but relieved. During the *kohabar* his wife's female kin introduced themselves in ways denoting the social proximity of their relationship.

Kichree

After a pause, the groom and his male kinsmen (I think they were mostly friends rather than relatives) entered the *maro*. At this point the drummers began playing again, and then continued for the rest of the time. The groom's mates added a lighthearted touch to the whole affair and were full of fun and jokes. The groom sat down again on the long bench with his "kinsmen" in a semicircle around him. Others, who were not taking an active part, stood around and mocked (in a friendly way).

First the groom, then his "kinsmen" were offered *kichree* (ritual food) to eat. They refused to eat it until satisfied with the gifts of money offered to them. The bride's father went first, putting money onto the groom's plate, followed by a smaller amount on the other plates. He was followed by his relatives, who did likewise, always starting with the groom, to whom the largest gift was made. To add fun to the occasion, the groom's friends vied with one another to see who could get the most. Finally, when the supply of gifts had dried up, the bride's father asked his new son-in-law whether he was satisfied. He was, and he tasted the *kichree*, as did his friends in their turn. Meanwhile, the groom's "father," having nibbled at the *kichree*, collected, then counted, the money.

Maro Hilai

In order to demonstrate his satisfaction with the gifts, the groom's "father" entered the *maro* and gently shook the central bamboo pole. At this, the bride's father handed him a small gift of money. The two shook hands and embraced. The groom's party immediately began to strip off some of the decorations as trophies.

The bride, who had changed into yet another outfit, went off to the sound of Indian film music with her husband and a chaperone. The newlyweds were to be welcomed in further ceremonies at his parental home, where she would spend three days before returning to her parents. The groom's mother had not been present at the wedding; she was at home preparing to receive the young couple. Later, after a few days back with her parents, the groom will come to fetch his bride and return with her to his parents' home, where they will have either a room or quarters of their own. The marriage can be consummated, probably on the Sunday night—a week after the wedding.

During the ceremony, the crowd of onlookers diminished. My assumption that people were beginning to drift home was completely wrong. They were all heading into the house for a meal of *roti* (unleavened bread), rice, pumpkin, curried beans, and a sweet made from crushed pea (or bean) fried in milk and rice, all served with soft drinks. Tables were set out in the area under the house, with spoons for those not wanting to eat with their fingers. We, as special guests, were invited to eat inside the house after the ceremony.

As the afternoon wore on, there were some non-Indians to be seen; but by the end, there were none. Throughout, one or two older women had tried to chant *birha* (wedding songs) in Hindi, but they soon gave up for lack of support. Hansar said that in earlier times the whole celebration would have been accompanied by the chanting of the women. One older woman, who tried to lead the singing, was wearing a big brass ring through her nostril and a heavy silver bangle on each ankle. Dressed all in white, she wore no shoes or sandals. I think she was the bride's grandmother.

Friday, April 10

Colin: *Matti Kore*

We attended the matti kore (planting of the marriage pole) celebration prior to the wedding of Ramnath Maharaj's daughter on Freeling Street. There were two pundits, one a brother of our host. Unfortunately, the all-women matti kore was just finishing as we arrived, though we were present for the subsequent *puja* and anointing of the bride.

During the evening we met Ramnarine Pundit, Ramnarine, the contractor, Mrs. Nandlal, Mistress Deabi, and Mrs. Rampersad.

Ramnath Maharaj told me that his daughter had been engaged to the groom western-style, though Ramnath had identified him in the first place. The groom works with Federation Chemicals as a loader at La Romaine.

Sunday, April 12

Gillian: *Ramnath Maharaj Brahmin wedding*

I made the following observations on the wedding of Ramnath Maharaj's daughter, which took place on Freeling Street, San Fernando. There were no drums. No preparations seem to have been made for the dwar puja. Rice was thrown over the groom before the *puja* began. There was nobody acting as younger brother of the bride.

This groom was dressed differently from the last wedding we attended. He wore tight-fitting satin trousers, gathered at the waist with a cord—like Mr. Bhattacharya. Over this he wore a creamy-pink, satin, knee-length, fitted coat with a pink chiffon pearl-studded turban.

At the milap, the supporters of the fathers of both bride and groom seemed to recognize each other well (Plate 5). It would appear that both families belong to the same social group. The lineup behind the bride's father was an impressive display of local wealth, power, and high caste. Among them were Mr. Binie and Mr. H. V. Gopaul.

The visor or veil, which at the Nandlal wedding was taken from the groom's headdress, was simply handed over by his party to the bride

Plate 5 *Milap* (meeting of the fathers of the bride and groom), Brahmin wedding, San Fernando. The bride's father—in glasses—and the *pundits* are wearing *dhoti* (loin cloth) and *kurta* (shirt). The Hindu spiritual leader of Trinidad, Jankieprasad Sharma, is carrying the black umbrella. In the background is the *maro* (marriage booth)

and her mother. It was pinned onto the heads of both mother and daughter by the bride's older married sister after the bride entered the *maro* for the wedding.

Both bride and groom were more relaxed than at the previous wedding. They looked at each other during the ceremony, and the groom spoke to his bride. On one occasion, he picked up her handkerchief for her. The bride did not sit on her father's knee. During the small ceremony performed by pundit and groom, the long bench (on which the couple sits as man and wife) had flowers strewn on it, and it was blessed. Mango leaves and rice were used in this ceremony. As far as we could see, no ball of flour and water containing coins and a hibiscus flower was handed over, though there was a short ceremony using a conch shell. The shell had a tall blade of grass rising from inside it; the interior was stuffed with crushed cotton and flowers. This shell was held in the right hand of the bride. First her mother, then her father, and the groom placed their right hand under that of the bride; then over the four hands the brother of the bride poured water, which was caught in a *tarriah* below.

The handing over of gifts by the groom's elder brother and his hanging of the pink garland around the bride's neck were all left to the end, after the sindoor dan. The materials he gave were marked; for example, there was a tab saying "to the person of the bride" and "for the *sindoor*." However, the gift labels were removed before the exchange.

The pundit, who was young and informal, spoke little Sanskrit during the ceremony, and frequently turned to greet friends among the witnesses. He even, at times, joined in with the singing of Indian film music that was being relayed over loudspeakers. A large number of pundits were present, at least seven of whom wore the *dhoti* and *kurta*. The Archpundit of Trinidad, Jankieprasad Sharma,[29]—an elderly man, who received a generous amount of money at the ceremony of the presentation of the *tarriah* and *lotah*—was also present. He wore a fine white turban and carried the inevitable black umbrella with a handle. He did not take his shoes off when he entered the *maro*.

At infrequent intervals, another pundit, Mahadeo, explained the ceremony over the microphone in English. He made comparisons with Christianity, commenting, for example, on the significance of fire and stone in the two religions (and in Islam). He translated the words of the bride and groom when they exchanged vows. The pundit said that the bride and groom would go round the fire seven times and that they would be husband and wife for seven incarnations. Later, Mahadeo told us privately that an important aspect of arranged marriages is the compatibility of the couple's horoscopes.

Although the passing of bride and groom seven times around the fire is supposed to be of the utmost importance in the Hindu marriage ceremony, this did not take place. The *lawa* from both homes was thrown over the fire certainly more than seven times, but on no occasion by the bride. As the *maro* was so crowded with people and objects, the pundit decided simply to pass a white handkerchief around the central pole himself, rather than have the bride and groom walk round it.

Much charcoal was burned throughout. The bride's nails were actually filed by the Nau when she first entered the *maro*, as were her mother's. The bride's eyes were decorated with red-and-white dots. At the end of the marriage ceremony, the young couple garlanded each other with flowers, and then they turned toward the witnesses and raised their hands in prayer before leaving the *maro* after touching the *shiva* stone. Once the marriage ceremony was completed, the bride and groom were greeted and anointed by close women friends of the family. This began with the bride's sister, who entered the *maro* bearing a *tarriah* containing two flour balls, some rice, and two gold bangles.

On this occasion, the presentation of the *lotah* and *tarriah* took place with great ceremony—far more so than at the Nandlal wedding. The hands of the givers were washed, and flowers and coins were put inside the *lotahs*. Before offering the gifts, each person washed the feet (or rather toes) of the young couple before touching his or her own brow.

This wedding was a display of wealth. Ramnath Maharaj does shift work for Shell and cannot be described as belonging in the upper income bracket, though we suspect that he is something of a property owner. The saris for the bride, her mother, and her sister had been brought especially from India. The *doolahin*'s (bride's) was a deep rich red—the traditional Indian bridal color—with quantities of gold thread finely worked into it, while her mother and sister wore similar green saris.

Two-thirds of the women present wore saris, and very fine ones at that. There were few flimsy, gaudy ones to be seen. It was above all a Brahmin wedding, attended by all the leading Brahmins. Young Mr. Rama Maharaj, whose wife does a lot of work in the *mandir*, was there. Mrs. Maharaj told us that their little daughter has pox, so they have stopped giving her meat, and will rub her down frequently with ghee. They must also keep the house spotless until she is better.

Bisram Gopie is apparently hoping to run for Naparima as PNM candidate. This is Seukeran's constituency. Mr. Gopie is hoping for Mr. Binie's backing. Mr. Treboransingh asked Bram why he has not been elected secretary of the Gandhi Service League, to which Bram retorted that he had no desire to be associated with a society that performs no useful function.

Why is Ramnath Maharaj able to mix with all these top Indians? From the social and financial point of view, he would appear to be below them. Is it caste? The same applies to Bram who moves freely among them. In Ramnath Maharaj's case, his high-caste status is matched by his spirituality.

There were few blacks present. Of the two policemen, one was black, the other Indian. There was plenty of vegetarian food or what non-Hindus call "Débé goat."[30] The signing of the marriage certificate took place in front of the wedding guests in the *maro* during the ceremony. According to Hansar, sometimes no marriage certificate is signed so that, if the marriage is a failure, it can more easily be dissolved. The groom gave his bride a wedding ring.

Thursday, April 23

Gillian: *Post-wedding events*

Eleven days after the Ramnath Maharaj wedding, we revisited the family on Freeling Street. Although the bride was back home with her parents, her husband had dropped in to see her. The pair had returned on Tuesday from a week's honeymoon in Tobago. The young husband will collect his wife from her parental home on Sunday, before taking her to his home for good.

Apparently, the newlyweds had gone to the groom's home for three days after the wedding, according to Hindu custom. Then, instead of returning to her parents' home, they flew off on honeymoon, after which the bride returned home. Consequently, the honeymoon has been neatly worked into the normal Hindu arrangement without disturbing it radically. The bride on no occasion wore either a white wedding dress or western going-away clothes. She is very fond of saris and wears them for all occasions. Her wedding sari was purchased by Mr. Bhattacharya in India. It is unusual in Trinidad to be married in a red sari.

Ramnath Maharaj has no *puja* room, but instead has constructed a shrine in a corner of the sitting room. There he has some pictures of deities, the Om[31] symbol, and a vase of artificial flowers. As a family they don't attend the *mandir* very often; the children go along, but not the parents. The parents are quite happy to practice their Hinduism in the home, as do many Hindus, but they go to the *mandir* to celebrate religious festivals such as Ram Naumi[32] and Diwali.[33] Mrs. Ramnath Maharaj celebrated Ram Naumi in the *mandir* at St. James, Port of Spain at midday on Monday, at the same time as Mistress Deabi and her friends were performing a *puja* in the *mandir* in San Fernando.

Gobardhan puja

In the evening we attended a Gobardhan puja, during a Bhagwat Yagya held at the home of Mr. Tindeo Jogie, who lives in Poui Avenue, Plaisance Village, Pointe-à-Pierre. Bhagwat is a colloquial term for a *yagna* (spelled *yagya* in Trinidad). This is a weeklong series of rites and ceremonial readings from the Bhagawata Purana—performed every day in the morning, afternoon, and evening.[34] The tent was a large galvanized-and-bamboo awning, with paper decorations including abstract patterns, milkmaids, and Mahabir (Hanuman). There were many *saddhus* present; they are men who renounce all material goods, and roam from one religious function to another. A great deal of Hindi was being spoken, particularly among the older people.

Under the canopy was a *vedi* on which the offerings had been placed. Religious pictures formed the backcloth—Hanuman, with garlands and flower decorations, Krishna and a cow, Shiva, and Ganesh. There were many little flags representing the planets of the solar system, and a light for the worship of Lakshmi. In front of the religious pictures, under an arch of small fairy lights, wrapped in red, yellow and white paper, was a position for the officiating pundit. All the participants were segregated, the pundits from the *saddhus*, and the women and children from the men.

The event opened with the garlanding of the officiating pundit, Karsiprasad Mishra, and some of the senior male participants, as a mark of respect. Women wearing long white dresses and veils walked around us offering fig (banana), *persad*, and melon. To the *saddhus* and beggars, they gave coins, in addition to the sacred food. We were told about one of the participating pundits, Mahadeo, a chemist at Brechin Castle sugar factory at Couva, who specializes in giving explanations in English on occasions such as this, then realized later that we had already come across him at the Ramnath Maharaj wedding. Mahadeo Pundit emphasized to the gathering that Sanathan Dharma stands for eternal righteousness or duty, and that there are four basic principles: truth, wisdom, mercy, and charity—a federation of other faiths. He referred to the *hawan*—fire—as "The Post-Master General of the Hindu faith."

Attention now concentrated on the Gobardhan mountain, made of mud, decorated with rice and red-and-white garlands, and linked from top to bottom by a spiral staircase. On top was a tall red flag. The pundit's assistant stood up and struck a *dhantal*, while the officiating pundit blew on a conch, and one of the *saddhus* played a wooden whistle. Chanting from the Bhagawata Purana began, with everyone joining in the refrain. An explanation of the text followed in Hindi, before the cycle was repeated.

Women circulated, keeping the *dias* burning by pouring *ghee* from Fernandes rum bottles. Bram defended the pundits to us, saying that he finds that they are frequently criticized too strongly. Would he say that in front of Hansar? Any attempt at dressing-up was confined to the pundits, the *saddhus*, and the family holding the celebration. Others were wearing working clothes. Wooden sandals with a cloth strap across were much in evidence this evening. We noticed that one beggar was wearing the *janeo*.

There was silence only when the pundit was reading or speaking. Otherwise, everything was accompanied by music, talking, movement—in short, confusion. The singing was led by a long-haired *saddhu*, who later told us that he had given up all worldly goods including his children and a big house. Later he had reclaimed his cows, but only to distribute them to the poor.

The finale to the *puja* was provided by the musicians, playing the flute, *manjira* (brass cymbals), *dhantal*, conch, and harmonium—a remarkable cacophony of sound guaranteed to reach the gods!

Saturday, April 25

Gillian: *Presbyterian fashion show in Penal*

At the invitation of Frank and Myrtle Cleghorn, we attended a meal, tea, and fashion show in aid of the Presbyterian Church in Penal. The atmosphere in the Cleghorn home in Penal was happy and welcoming, and we were given a delicious lunch of traditional Indian food—rice, curried chicken, *dal*, *dalpuri*, and salad. Knives and forks were laid, but Frank, Colin, and Mr. Gafour (Myrtle's uncle) ate with their hands. The women used a knife and fork.

Myrtle Cleghorn, whose maiden name was Khan, is of Muslim origin. Her husband, Frank, was born in La Romaine and owns two lots on the Southern Main Road leading south out of San Fernando. Myrtle's father is an inspector of schools and was formerly headmaster of the Presbyterian School at Penal. The Cleghorns live in the Penal School House, and Myrtle is president of the Women's Mission. They appear to be pillars of the Presbyterian Indian community.

Frank is 32, and Myrtle must be the same or younger. They have been married for eight years, and have five children, of whom the eldest is seven.

Colin

Mr. Gafour is Muslim, and his mother is a sister of Myrtle's father. He has been married for 14 years to a light-skinned Barbadian. His wife is Anglican, but their eldest child, aged ten, is Catholic. Despite being Christian, she attends Muslim primary school, can read the Koran in Arabic, and is beginning to learn Urdu. Mr. Gafour works for a firm of contractors. He says there is no caste among Muslims, though there is some ascriptive ranking, such as that involving the Khans. He claims that he is welcome at Hindu ceremonies, such as Bhaghwats, and boasts that he has more close Hindu friends than of any other religion.

Gillian

The tea party was opened formally by Colin, though there were few there to hear. Tea was served in china pots, and there were dainty sandwiches, cakes, and meat pies. The Canadian Mission School provided the venue, and there was a good atmosphere with music blaring through giant amplifiers. Here Indian film music was replaced by Mantovani[35] and his orchestra.

The women were smartly dressed. Almost all were East Indian, though western in appearance, with short, wavy hair. The models had first names such as Christendaye, Myrna, and Lystra, and family names such as Baldeo, Ramkissoon, Rampath, Rampersad, and Maharaj. Cleghorn was by far the most English-sounding name there. Most, if not all, the models were Presbyterian.

About 40–50 people attended the tea, and approximately 150 to 200 watched the fashion show (excluding models and helpers). The relationships between boys and girls there seemed quite westernized. The boys adopted a sexy approach to the models and their figure-hugging sportswear. When opening the fashion show, Myrtle asked us to sit for a moment in silent prayer, as the minister was not there.

Myrtle's two sisters are still at school and attend St. Mary's Convent. One wants to be an air hostess (though this is usually thought of as a Creole ambition). Myrtle's father owns a big American car, and is an elder of the Marabella Presbyterian Church. Her brother is a legal adviser at Texaco, Pointe-à-Pierre.

Sunday, April 26

Gillian: *Temple foundation stone, Pepper Village*

Today Bram, Colin, and I went to Pepper Village, just outside Fyzabad, to witness the laying of the foundation stone for a Hindu temple by Bhadase Maraj, who had not turned up by dusk when we left. The carefully prepared foundation stone was left lying abandoned in a corner of the site. The main feature of the afternoon was a bazaar, with melon, candies, channa, nuts, coconuts, and masses of soft drinks for sale. In a palm-leaf tent, you could even buy traditional English teas, consisting of a cup of tea with two sandwiches and a piece of cake. As we arrived, a young girl in a *sari* came up and pinned a tiny pink paper rosebud onto each visitor.

Music was provided by two rival groups. The first was an Indian band playing Indian film music with the odd calypso thrown in for good measure. The second group, which aroused much attention, was The Maharaj Kids from San Juan, a steel band, consisting of four brothers and a sister. They specialize in pan arrangements of Indian film music, calypsos, and English music. Immensely popular, they were surrounded by a large crowd whenever they played, but it was noticeable that people preferred to stand around and listen rather than "jump-up" as a Creole audience would do.

Most of our time at Pepper Village was spent eating, drinking, and talking. Soft drinks were served, though a house nearby was being used as a bar for those who preferred the hard stuff. This was well frequented throughout.

I discussed Hinduism with an Arya Samaji from California (Trinidad). He told me that Arya Samaj represents the new, revolutionary approach to Hinduism. Sanathan Dharma is for the Brahmins, but the Arya Samaj welcomes all lower castes, particularly the untouchables—by which I think he meant the very low castes, since all Hindus had made themselves untouchable by crossing the *kala pani* (black water or sea) to Trinidad. In his opinion the true spirit of Hinduism is to be found in this group and in their teachings. He added that there were several groups in San Fernando, but there is some dispute over leadership at present. Norman Girwar is leader of one of the groups. Arya Samaj aims to try to reach people in English, and above all explain the significance of the ritual, and the meaning of the sacred books. My informant is much influenced by Dr. Radhakrishnan[36] and Mr. Bhattacharya. He feels that the pundits cannot give the spiritual lead they should. They explain little or nothing, and frequently what they do explain is not correct.

The Arya Samajist reckons that Hindus from all castes except perhaps Brahmins went over to Christianity. He does not think that it was largely low-caste Indians who converted. He finds many Christian Indians lacking in what he would call culture. In other words, having abandoned their own oriental religion and culture, they adopt the outward manifestations of western culture. In his opinion, Indians moved over to Christianity mostly in order to improve their own social position. The Canadian Mission gave them the chance of education, more money, and higher status in society. And this was the bait. In fact, it was the way out of the cane fields.

Later I had a conversation about cremation with Miss Elektra Chandroo, a Presbyterian. She and her women friends have all attended cremations at Mosquito Creek, south of San Fernando, but she has no idea what really happens at this ceremony. She intends to leave a note with Mr. Bhattacharya before she dies, requesting that her body be cremated.

The crowd there was predominantly Indian, but from time to time groups of blacks arrived and mingled quite happily. Dress was informal. I think that Mr. Binie was one of the few men in a suit. A handful wore the *sari*. There was a surprising atmosphere of gaiety in the place, and I imagine it must have been a great success—even though no foundation stone was actually laid.

Colin

Mr. Gopaul was there, together with Binie Maharaj, Mr. Binie's wife, and granddaughter (she goes to the temple on Sunday mornings).

Bram and Mr. Bholasingh (from California) had an interesting discussion about caste. Bholasingh is a Chattri (Kshatriya). He has a much stricter interpretation of caste than Bram. Bram's view is that all men must do their duty. Bholasingh is very concerned with the ritual impurity of non-Brahmins—including himself—and is critical even of Hansar.

There was much criticism of Bisram Gopie, who is, according to Bram, "a Chamar who does Chamar's work." Bholasingh says that he has been associated with Gopie in prayer meetings and that Gopie does not conduct himself well (that is like a Brahmin). He praises him for his social work—it is presumably because of this that Gopie has been so closely associated with Binie Maharaj in the temple and the Gandhi Service League.

The Todd Street temple in San Fernando is not associated with the Maha Sabha. It seems to be in the pocket of Mr. Binie and was formed by him, a Mr. Seunarine of Siparia, and Mr. Gopie. Gopie is favorably disposed toward Bhadase Maraj,[37] because the latter is drawn to the PNM. Mr. Binie and Mr. Gopaul are associated with the Gandhi Service League. Both wish to be remembered for their piety. Mr. Gopaul is thinking of founding a Hindu seminary.

Bram collected 300 signatures petitioning for a Hindu School—and got multiracial support. It was approved by the Ministry and the Maha Sabha—but not supported by Mr. Seunarine and Mr. Gopie, who said that the temple site could only be used for a *siwala*. Bram thinks they opposed it because it was not their scheme.

Bram would prefer his oldest daughter, Tara, not to find a job when she leaves school. He realizes, however, that this is not what she will want.

Monday, April 27

Colin: *At the San Fernando probation office*

I went to the Probation Office with a Probation Officer, Mr. Bullock, and Frank Cleghorn. The chief officer is Mr. Manickchand, and the entire office is East Indian. Mr. Bullock had some interesting stories about their marriage-guidance work, which is mostly with East Indians, some because of arranged marriages, others because East Indian women are "old-fashioned"—for example, in dress. Most marriages are still within caste, and this applies not just to Brahmins.

There is sometimes intermarriage between Hindu and Presbyterian East Indians. Bullock's cousin was a "good match," and she married a Hindu boy by Presbyterian rites. When they reached the boy's home, a traditional Hindu *puja* was held. Presbyterian East Indians have maintained many elements of Hindu and Muslim lifestyles: taboo on beef and pork (often broken when abroad); a strong parental hand in deciding marriage partners; and protection of daughters.

Frank told me about his father-in-law, Mr. Khan[38] of Marabella, and his fondness for Frank's eldest boy. On the day after his birth, Frank's son was taken to his grandfather, who at that time lived at Sangre Grande, and he spent his first three months there.

By sitting with Myrtle, Frank is the odd man out in church, because gender segregation is the norm in rural Presbyterian congregations, though this is not the case in San Fernando. Until ten years ago, a regular Presbyterian service was held in Hindi at Susamachar (the Church of Glad Tidings) on Coffee Street in San Fernando.

Later Frank and I went on to Débé, Penal, and Siparia. I met the heads of several Presbyterian schools—at Penal and Débé (Mr. Rajkumar Maharaj), and at the Hindu School at Débé (Mr. Mhunital). Francis Seupaul and Philip Sukal of the Presbyterian School at Débé have offered to help me with my survey of the village.

Tuesday, April 28

Gillian: *Dr. Stella Abidh, Medical Officer of Health*

We had a conversation with Dr. Stella Abidh,[39] medical officer of health for San Fernando. She told us that there is no typhoid fever and no polio in the shantytown, and that there is practically no yellow fever or malaria in Trinidad. The lack of disease in Port of Spain and San Fernando shantytowns presents a problem—if they were disease areas, the government could bulldoze them at once. Politicians are keen on keeping shantytowns because they are important for votes. Dr. Abidh advised us to be taken into shantytown by a black health officer—a peculiar recommendation, since the shantytown is a major East Indian enclave.

Why are there Indians in the San Fernando shantytown? According to Dr. Abidh (herself a Presbyterian Indian of Muslim origin), the Indians are really too proud to squat anywhere. But they have observed that black squatters get government help in the form of housing schemes, so they want to share that benefit.

Dr. Abidh is intensely aware of the link between race and politics here. She says that 10 to 15 years ago there were Indian civil servants, and Indians were to be found in the banks. She claims that blacks alone are now favored for positions in banks—and that in crowded classrooms in schools, Indian children are pushed to the back, which puts them at a disadvantage. She observes that now "black is up, he'll do all he can to prove he was given a raw deal by the British, and therefore does everything possible to assert superiority." There is a great deal of scheming to get votes. The government has plans to expand black housing where villages already exist as East Indian communities. This is done simply to gain PNM voters.

Indians are becoming keener on family planning, whereas blacks are not. Dr. Abidh predicts that the Indian birth rate will not continue to increase as it is at present. She claims that the Indian family unit is more open to family planning. Black women, with several fathers to their children, are unlikely to "plan." There was much vehement opposition to the Family Planning Clinic on Lord Street, San Fernando from Roman Catholics. Hindus and Muslims also oppose family planning.

Dr. Abidh thinks that Trinidad is "communist inclined,"[40] and cites the links that Williams has made with Yugoslavia and the Arab countries, the influence of Jagan in British Guiana, and the indifference of Downing Street. She repeated Nkrumah's slogan, "Seek ye first the political kingdom," and referred to the fixing of voting machines in the last (1961) election.

Débé Canadian Mission Indian (CMI) school had been recommended for demolition before it burnt down. Already the Canadians have raised one-third of the cost of rebuilding, which, as the denomination concerned, it is required to pay. Years later, there is still no new building, because the government has failed to supply the other two-thirds (and the children are being taught in the market building).

Dr. Abidh said that she intends to introduce us to the Balkissoons—a highly intelligent, poor Brahmin family. A pundit plays an important role in their spiritual life, and in all other aspects of family life. The boys are in the professions; one daughter is a nurse at San Fernando hospital; and the second is a teacher. Yet they are very devout Hindus. She also wanted us to meet the Ajodah Singh family—the father was the model for Naipaul's *The Mystic Masseur*.[41]

Dr. Abidh claims that low-caste Hindus are frequently attracted to Roman Catholicism because of the ritual (note importance of ritual in Hinduism). Siparu Mai is associated by Hindus with Kali Mai.[42] According to Dr. Abidh, black virgins are always in some way connected with water: they have either crossed it or are found in it. A black virgin is never found carrying the child Jesus. Siparu Mai is carrying a staff in one hand and a chaplet in the other. Kali Mai carries a sword.

Food taboos persist among Christian Indians, and are only broken by those who go away to study. Many Christian Indians still follow the beliefs and superstitions of their grandparents. Arranged marriages are common among them. In other words, much of the choice of partner lies with the parents.

Colin

Frank Cleghorn says that his father's father was half black. He claims that he has many black friends. He does not object to a black man dancing with his wife and putting his arm around her.

Thursday, April 30

Gillian: *Hansar, Rosalind, and Bram*

Hansar, Rosalind, and Bram came to our house for a talk. According to Hansar, the term *saddhu* in Trinidad is synonymous with beggar—very different from the true Hindu meaning.

Rosalind gave us some kinship terms:

Father's father = *dada*
Father's mother = *dadi*
Mother's father = *nana*
Mother's mother = *nani*

Ma and Pa are used for parents; she says she knows and uses no other terms for relations, such as father's elder brother.

Hansar believes in planetary influences on one's life. When you want to consult a *patra* or horoscope, you must approach a pundit; 80 to 90 percent of Hindus in San Fernando will not consult a pundit as regularly as they should. They only do so if they want to enter business or win money on horses.

The idea behind a father first seeing his child by reflection in a mirror is to absorb his power to damage the child. Black "ash" is put on women's faces and on children to attract attention, so that they cannot later be harmed by *maljeu* (evil eye).[43] Black beads—"*jumbie*[44] beads"—are put around the wrist of a child to ward off evil spirits. They are worn from shortly after birth to one year of age. Hansar did this for all his babies. Black was put around the eyes of the oldest ones—now he is "more modern." *Kardhan* is a string, tied around a child from two to three weeks after birth. It is a colored thread, firmly knotted, and put round the bottom of the belly to prevent ruptures. Only the oldest of Hansar's children had this.

Barahi is a celebration that takes place 12 days after a birth. Long ago it was a big celebration to which the whole village was invited; it is still observed, but it is a smaller event now, with more drink. Hansar says it was easier in former days, because there was no need for knives, forks, or plates. Now people insist on having these to show their status, and the result is they cannot invite so many people. Hansar thinks that things are changing in the country and in towns. Now people are more conservative, more money-minded. Indians, as a group, are mocked for sitting on the ground, and eating from a leaf with their fingers.

After puberty girls should be kept at home to avoid the risks of exposure to society through education and travel to school.

There has been an increase in parent/child conflict. Hansar does not think that parents can nowadays choose a spouse for their sons and daughters. There is no longer the close contact with children that you find in villages. All is fine in a village context: Hansar notes, "our problem is a social problem."

When asked about racial intermarriage, Hansar commented, "What would happen if we produced a whole nation like our problem child, Eric Williams?"[45] A neat way of worming out of a dogmatic reply! Hansar thinks that it is only men with no family of their own who suggest racial intermarriage. Hansar doesn't take newspapers; Bram reads *The Mirror* carefully.

Sunday, May 3

Colin: *Ena Scott-Jack on "social" Indians*

Discussion with Ena Scott-Jack revealed that there seem to be few "social" East Indian families in San Fernando, for example, like the Wilsons, Sinanans, Kangaloos, and Namsoos. Bisram Gopie is quite well regarded among Creoles (because of his social work and also, I suspect, because of his PNM commitment). Mr. Dialdas is also respected by Creoles. He donates prizes for the San Fernando Carnival.

Mr. Gopaul and Mr. Binie are not "social" East Indians, though Ena does not rule out the possibility that Gopaul may be accepted (Mrs. Roxburgh held a similar view about the Port of Spain Hindu, Jang Bahadoorsingh).

Creole and East Indian ranking systems do not mesh, in part because of the separate issues of color and caste.

Andrew Carr on the PNM

In Port of Spain I talked to Hans Guggenheim[46] and Andrew Carr.[47] Hans, who has been to India, says that much of the pundits' ritual is suspect (which reinforces Hansar's opinion). Andrew Carr, public relations secretary for the PNM, knows little about East Indians, caste, or Indian politics. He notes that the great problem is to encourage PNM party groups to discuss issues out of election time. Andrew knows Mr. Gopie, who, he says, is trying to penetrate the Hindu wall. Andrew also knows about the Gandhi Service League, and describes Gopie as dedicated to the type of rural development outlined by Gandhi. He referred to the difficulty of persuading blacks that not all whites and East Indians are anti-PNM.

Monday, May 4

Colin: *Organized Hinduism and caste*

I recorded the following entries made on loose paper during early May.

Bram told us that Bisram Gopie is the president of the Gandhi Service League; Binie Maharaj is secretary and H. V. Gopaul, treasurer. Mr. Gopie's view is that "Service to Man is Service to God." They hold no regular meetings. Bram is not always invited: Hansar is invited but is not a member. Mr. Gopie usually gets a speaker and invites one or two PNM councilors in order to impress them that he controls the Hindu community. The *siwala* stands for "truth, love and nonviolence" but has no legal committee. Mr. Bhattacharya was invited to conduct the Gita class by Hansar and a few others, but very few young Brahmins attend. One pundit, Vishnudath, an Arya Samaji from Diamond, regularly participates. About a quarter of those in the Gita class are Christian.

Mr. Bhattacharya's most recent Gita class, which we attended, was largely devoted to a comparison between the caste system and feudalism. The name Maharaj was adopted from Brahmin rulers. Chamars were likened by Mr. Bhattacharya to Mary Magdalene. *Gotra* is based on shared descent, as in the case of Mr. Bhattacharya and his father and grandfather. One's last thoughts at death are pointers to the nature of one's next incarnation. It is clear that Bram and Hansar consider that Mr. Bhattacharya is more intelligent than the black, Dr. Baldwin George, whom we (eventually) heard give a talk about the Gita. However, they were pleased at the interest Dr. George had taken in Hinduism.

Hansar told us a complicated story about Bissoon Pundit and his letter to the Gandhi Service League, complaining about his (Hansar's) taking the Sunday temple service, because he is not a Brahmin. He also told us about Binie Maharaj's black children. Hansar concurs with caste only as it affects ritual status. Brahmins are the traditional priests; but they are ignorant, and others who are not ignorant should not be excluded from the priesthood.

Wednesday, May 13

Gillian: *Future of Indians and Creoles*

I held a conversation with George Sammy, Peter Dubé, and Duffy Mohammed on the future of Indians and Creoles:

Will East Indians and Creoles eventually merge?

(George) No. On the street, yes, and apparently in homes. But in fact much will be retained in the homes. There is a tendency for East Indians to go western in youth, eastern in middle age.

(Peter) The middleman will retain Indian culture; those both at the top and bottom of the social scale will become westernized.

(George) Despite his wife's reservations, he gave each of his children an Indian first name.[48] George sees the increase in the number of temples in Trinidad as proof that a conscious effort is being made to retain Indianness. The same may be said of Hindu and Muslim schools.

Peter regards all this as a last-ditch stand. Before the founding of the Sanathan Dharma Maha Sabha in 1952, Indianness was at its lowest ebb. There was no sense of belonging to an Indian community: there was no Indian community. People felt ashamed to speak Hindi and wear the *oronhi*, because of their "coolie" associations. But now the sari is being worn, for example, by Mrs. Ruth Seukeran[49] and her daughters.

(George) His parents realized that, if their children were to get on socially and economically, they would have to become Christian and western. George's wife is distanced from her Indian background. She is third- or fourth-generation Trinidadian; George is first generation—his parents came from Madras.

(Peter) Indian films and radio programs in Trinidad are introducing Indianness, rather than encouraging the retention of Hinduism or Islam.

(George and Peter) Muslims have established themselves strongly in Trinidad. Eastern influence on the west in Trinidad is manifest in the sari and *roti*.

(George) There will be a merger in future (modifying his first opinion). Much that is Indian will be lost, but Indians will always retain a certain identity. They will take their place in society and yet stand apart. They will not completely abandon their culture as blacks are perceived by the Indians to have done.

Questioned about intermarriage (between blacks and Indians), George replied that it is certainly taking place nowadays.

Peter explained that the reasons were social and economic. The upper classes are tending to mix legally (through marriage); the lower two-thirds are likely to have sex but not marry, therefore forming common-law unions.

(George) Intermarriage is easier now than it was formerly. When he was a boy, the whole village would have been up in arms; now it is not condoned, but reaction to it is less vehement. About 60 percent of colored lecturers at UWI (Creoles and Indians) and about 50 percent of doctors and lawyers in Trinidad are married to Europeans. [50]

Indians and blacks do intermarry, but are generally not accepted by the Indian family. Relatives of George's wife who married blacks have been ostracized by the family. Nevertheless, once an Indian/black marriage has taken place in a family, it is much easier for it to happen a second or third time. Myrtle's cousins who intermarried were all Indian girls who married black professional men. They were Presbyterians.

One Indian girl in George's family returned from university married to a black. When her family came to visit them, the black husband retired to the back of the house. She had left home in order to marry him. Indians will marry Chinese or Europeans, but not blacks. Reason—"black has no culture."

Why do Indian girls marry black men?

(Peter) It gives them an opportunity to get what they want—a western way of life.

(George) The reasons for intermarriage are (a) a sense of adventure (this applies to a small number of people) and; (b) escape from one's own group when one is not really accepted in it.

(Peter) Economic factors are important; to the black, intermarriage is a step up in the socioeconomic ladder. In the eyes of the Indian, the black is part of the west. Therefore, if young Indians want to identify with the west, they will ipso facto mix with blacks.

Hinduism is more likely to survive now that books on the subject are available in English. Young Indians will read first in English, then become aware of the inadequacy of the English translation, and want to learn Sanskrit and Hindi in order to discover the truth.

(Duffy) He has two male cousins with illegitimate children from black women. Both these men later married Indian women. The children live with their mothers and are not regarded as members of the Indian families.

General conversation followed

(Peter) From Bhadase Maraj's evil, good came—Hindu schools and temples.[51] Dr. Abidh's father was murdered. He, as a Presbyterian, opposed the founding of Hindu schools by Bhadase. He played on Indian factionalism in order to prevent it happening. In Hansard, Ranjit Kumar is quoted as saying, "The ghost of Abidh[52] will always haunt you."

For Bhadase Maraj, power is money. Anamunthudo is in charge of the Vedic Press. He is a strong socialist and disciple of C. L. R.

James.[53] According to Peter, "Anamunthudo is politically twisted." For Norman Girwar, too, power is money. George has no respect for him. Girwar is quoted as saying, "Let's get into sugar. Sugar is money." Girwar claims, "I am a socialist and will have nothing to do with any political party until they declare themselves socialist." But George says that he knows that Girwar is, behind the scenes, playing a government game. Girwar's father-in-law was a cane farmer.

Factionalism is rife among East Indians, particularly among the leaders. Peter comments that Tableland is notorious for East Indian family feuds lasting generations. Tableland is commonly called "no-man's-land"; there have been fantastic burnings and shootings there in the past.

Peter says that in some cases a translated version of kinship terms exists in Presbyterian Indian homes: for example, big aunty (*barki mai*). George's elder sister used to be called *didi*, shortened to *di*. She is now known as *Di* in her village, as if it were her first name. Peter's relations change *phuphu* (aunty) to *phup*, hence "Anand *phup*" (Anand's aunty). Duffy's mother always wanted him to marry a Presbyterian Indian girl. He has never married. Although his last name is Mohammed, his father was a Hindu who converted to the Presbyterian Church.

(Duffy and Peter) Mechanical jobs were formerly taken only by blacks—in fact, they carried out all the work in the oilfields. Indians believed they were not strong enough for these tasks. But now there is infiltration by Indians. Scavenging and water carrying were formerly Indian preserves; now more blacks are involved in these tasks. In the past, no Indian would be a soldier or a policeman.

(Peter) It was rumored that police had to take an oath to arrest their own mother if necessary. Also diet was a problem, such as the food in police canteens. Formerly, there were only black postmen. Now there are more Indians, but the pay is too poor to attract Indians.

(George) Indians were formerly ranked in society as individuals, not as members of the Indian community. In the top bracket were Fitzpatrick and Grant (both Madrasi). They had this social position, but they were in no way associated with the Indian community. Charlie Fitzpatrick would dine with the governor in the week and spend the weekend swigging rum with the villagers in Duncan Village.[54]

(Peter and George) Indians in Trinidad have a strong affinity with India. They admire Indian culture with all its mystery and learning. But they have no desire to go back to India to live—simply a desire to maintain ties.

Names of prominent San Fernando Indians: Ralph Soodeen;[55] Daisy Roodal; Dopson (San Fernando family with "Indian" blood); Appleton. All Clayton (*dougla*) Appleton's brothers and sisters married black/whites, that is, Creoles, except Clayton. His parents were never identified as East Indian; and were always out of Indian society. Eileen, Clayton's wife, is a very light Indian. Eileen and Clayton appear to mix entirely with Creoles.

Thursday, May 14

Gillian: *A similar discussion with Bram and Hansar*

In conversation with Bram and Hansar, Bram commented that there will be a total merger of Indians and blacks in the future. His conclusion was based on two lines of argument. The Maha Sabha is failing in its duty; for example, first, there should be at least one person to teach Hinduism in Hindu schools. Second, temples will become gambling dens if they continue to fail to have something to offer to society.

Blacks want integration. Bram would ideally like Trinidad to be divided into two equal parts for Indians and blacks.[56] He is sufficiently realistic to stress at once that this would never be possible. Bram claims that the PNM is at present employing Indian girls among black men to encourage integration.[57] Before 1956 there was discrimination, but it was "like a book on a shelf that nobody bothered to look at." Once it had been opened in 1956, everyone became eager to find out what it was all about. Since 1956, race has been used in all elections, by Eric Williams, in particular.

Having Indians in the PNM is "eyewash for public consumption." Kammaludin Mohammed[58] and Dr. Winston Mahabir[59] have both completely given up their Indian culture. They are only Indian by birth. When canvassing for the DLP in 1961, Bram often found his reception in PNM homes very poor. They were unwilling even to hear the DLP point of view.

Conversion to Christianity among Indian youth is taking place rapidly, but Bram sees it as a conversion to a western way of life, with little or nothing to do with religion. Older Presbyterians have frequently retained much of their Indianness. Bram considers that this Indianness means nothing to young Presbyterians. Bram says that you are likely to find segregation of the sexes in any Presbyterian church in country areas where there is an East Indian majority. As soon as they are in a minority, you will find them adopting western customs.

The actual ceremony performed at a Hindu funeral is the same as that at a cremation. The burial ceremony was formerly much longer. The widow is, strictly speaking, regarded as half-dead. The red in the parting of her hair used to be removed and ash put in its place. This is not done nowadays. The eldest son leads the children five times around the grave. At each turn, a little earth is picked up. Stopping at each turn is a symbolic way of saying farewell to the various material aspects of the dead person. Each stop marks the return of one element to its rightful place.

We are told by Hansar that there are 14 Selfless Service Divine Mission groups (once there were 17, but some split off, as they thought they could manage on their own, only later to break up). Each group holds a Sunday service, but the problem is to find leaders to keep the groups going (Plate 6). Princes Town has the furthest group from San Fernando. There are also groups at Gasparillo (2), Reform, Golconda, and La Romaine. The second Princes Town group is folding because the teacher who ran it has been sent elsewhere in the island, and there is no one left to take his place.

On the future of Indian culture, Hansar thought there would be a fair amount of East Indian and black merger. But it would never be complete, because of growing self-consciousness among both Indians

Plate 6 Hansar Ramsamooj (in white hat) resting during the singing of *bhajans* (hymns)

and blacks. Moreover, Christian converts were vigilant to maintain Indian culture. The Indian way of life is already a blend between east and west, and therefore a compromise. What Hansar and Bram thought of as Indian culture is fast disappearing in India, too. Hansar stressed the importance of Indian films, and of visiting *swamis* and lecturers in bolstering Indian culture in Trinidad.

Hansar commented that many Trinidad blacks accept Hindu culture, as do Americans and Europeans. He regards Hinduism, the religion and the way of life, as Indian culture. He is more optimistic about the future of Hinduism than Bram. He underlined the importance of the time in which we live. Blacks are now in the ascendant, and have a great desire to be respected. "They have always been the underdogs—one can't blame them if they lord it a bit."

Bram added that the advent of the PNM in 1956 has done good, in that there has been an awakening among Christian East Indians about their cultural heritage. They are now eager to find out about India. This new awareness has come with the development of the idea of an Indian community. The PNM's playing on race makes the Christian East Indians feel as if they are "trespassing in the black camp." Now the Christian East Indian has plenty of pride and wants an identity. In Bram's view, Eric Williams can't be impartial. If he starts treating Indians well, all those who back him will let him fall. He is power crazy.

Discussion moved on to the traditional method of selecting a bride. Once a young woman has been identified,[60] the boy's father goes first to a neighbor's house, and if the neighbor is not complimentary, he goes away. If the neighbor responds favorably, the boy's father approaches the girl's home, looking carefully at the yard, the appearance of the house, both outside and in. The two fathers meet, and the visitor asks to see the girl. Her father makes an excuse, such as a request for water to be brought, so that she enters the room without knowing what is happening. Once the girl has been approved, there is very little talk; when talking begins, it is mostly about money.

We also discussed the divine nature of the mother figure. Mothers are always of paramount importance to Indians. The mother rubs down the child from birth, and shapes its face with oil. Hansar's mother rubs him down even now when he is unwell. A large family enables household duties to be shared, though there is a current tendency for daughters to go away for professional studies. Hansar adds that education equals security.

Sunday, May 31

Colin: *Conversations in Chaguanas*

We went to Chaguanas with Bram and his daughter Tara, and met Bholasingh and his Arya Samaji friend from California. We were soon introduced to Brahmachari Hari Ram, leader of the Divine Life Society of Trinidad and Tobago, and the country's only Hindu monk. He travels around the country preaching, teaching, and holding *hawan*. He was originally at Felicity[61] and is now at the St. James *mandir* in Port of Spain. The latter is sponsored by Jang Bahadoorsingh; most temples and the functions held in them seem to be in the pocket of a boss or headman.

Bholasingh thinks that nowadays not many East Indian marriages are arranged, and they are split fifty-fifty within and outside caste. It all depends on the control exercised by the parents. Bholasingh's uncle married his first three sons to Chattri girls. Bholasingh thinks that, if he does marry, he should marry a low-caste girl to prove that caste is not important to him. Bholasingh comes across as strongly anti-Brahmin, objecting, in particular, to their stranglehold on the Maha Sabha. He thinks that a Chamar, Dr. Avatar, should be made head. Yet, only a few weeks ago, Bholasingh told Bram that he thought that priests should be Brahmins, whereas Bram told a Chamar acquaintance that caste should die out. Attitudes to caste are ambivalent, and shift with the person to whom an individual is speaking.

Bholasingh claims that most people stick to caste for reasons of snobbery and tradition. It really does not have theological significance for them. It is the Brahmins and Chattris, especially, who marry within caste. The only reason why a Brahmin in Débé would agree to his son marrying a Sudra girl would be that she was very wealthy; and she would be more acceptable to him than a poor Brahmin girl. He concluded that Hindu piety is fast disappearing in Trinidad—and most rapidly in the towns.

Later in the afternoon, we met Mr. Mahadeo, a relation of Bram's who is on the Executive of the Maha Sabha. He told us that there are about 37 Maha Sabha primary schools and 2 secondary schools (co-ed at Sangre Grande, and girls only at the Maha Sabha headquarters). Hindi and Sanskrit can be taken at GCE "O" level, if needed. We also talked to Mrs. Kalkar, a lecturer in Sanskrit from the Central Provinces of India. She teaches a Gita class at Curepe and gives lessons in Sanskrit and Hindi. The Sanskrit students are mostly non-Brahmins who cannot become priests. Bholasingh insisted on introducing us to a *Trinidad Guardian* reporter, Herman Roop Dass, who will, I hope, give us a fair write-up.

We met an imam from California and his brother, Mohammed, from Chaguanas. They told us that it is the practice for the imam to be elected by the congregation; he is usually middle-aged, and the most knowledgeable and respected man in the community. Muslim customs with regard to food, women, and the upbringing of daughters and choosing of brides are very similar to those of Hindus. Girls are still wed by proxy, but there is now a great deal of choice within the age-range 19–21. Very few Muslims converted to Christianity, the major exceptions being members of the teaching profession.

In the temple in San Fernando this morning, Hansar paid tribute to the Indian leader, Nehru. But he warned that India was progressing rapidly, and was leaving behind the India that existed in popular imagination. India had made great steps forward and had abolished the legal status of caste.

Tuesday, June 2

Gillian: *Focus group discussions*

George Sammy had invited a number of people to his home in Westwood Street to form a focus group. It was proposed to debate a number of issues associated with East Indian life, so that their views could inform the questionnaire we were in the process of preparing. Present were: Bram, Mr. James Sammy[62] (formerly a senior teacher at Naparima College), Mr. George Sammy, Mr. Peter Dubé, Dr. Makhan Dubé[63] and Mrs. Vilma Dubé, Mr. Roy Mootoo, Mr. Sookoo, Mr. Bhattacharya, Mr. Mootoo (Sr.), Dr. Ramesar, Mrs. Ivy Gunness, Mr. Sayed Hosein.

The following questions were put to the group by Colin:

Is there such a thing as an East Indian Community?

James Sammy commented that there is no solid East Indian community in San Fernando. Such cohesion as there is, lies one-third in racial origin and two-thirds in religious background. Their weakness lies in the fact that as a group they are not solid, and lack cohesion—such as clubs and social organizations. But much is being done by each religious group. In earlier times, the main activity was in the Christian Church, but now there is a revival of interest in Hinduism and Islam in the context of renewed political and social awareness.

George Sammy added that Oriental Cricket Club and Oxford Club overlooked religious difference. Oriental has been established for 50 years, yet it has no clubroom or ground. But these clubs are more for the young, and involve comparatively small numbers. James Sammy noted that these two clubs are primarily for dancing and sport; no real social work is being done through them. There should be more of an effort to raise morals.

Vilma Dubé wondered why the Indians have produced no outstanding leaders with a social conscience. They have been too interested in power seeking, and in using the church or the temple for their own prestige. In Vilma's view, the East Indian community is fragmented along religious lines, but with a tendency toward cohesion. James Sammy pointed to the religiosity of Indians, but Mr. Bhattacharya added that Hindu teaching in India was by example rather than precept.

Indians were on the lowest rung of the economic ladder in Trinidad, James Sammy observed, and therefore sought material gain in a big way. At one time, Muslims, Hindus, and Presbyterians were all one in the Presbyterian Schools, and many even attended church services. It was the arrival of religious leaders from India—in the 1920s and 1930s—that drove them apart.

George Sammy commented that education cemented the three groups, but James Sammy came back with the view that education and the social groups clustered around the Presbyterian schools, and that no alternative was provided by the other religions. Nobody except the Presbyterian Church bothered about both the educational and spiritual needs of the Indians. George Sammy added that there was a slow infusion of the Christian spirit into primary education.

So, why convert?

Presbyterian schools always had a secular and a religious function. Why convert in the present day? Makhan Dubé pointed out that in Princes Town, fewer than ten (perhaps between five and eight) had converted in the last year.

Social disabilities, such as caste and untouchability, were often the reason for conversion, according to Mr. Bhattacharya. In British Guiana, lower castes sought equality in conversion. He cited the case of a staunch Hindu of low caste who was with a group of Hindus who refused to eat with him. In anger he turned to Christianity in order to be acceptable—and regretted it later. His sons were Christian, but retained a deep and scholarly interest in Hinduism. Sometimes a social misdemeanor or marriage problems might lie behind conversion.

Makhan Dubé emphasized the escapism aspect, plus the social and economic advantage brought by conversion.

James Sammy did not agree that only the low castes converted to Christianity. He knows that all castes are represented in the Presbyterian Church. It is only to be expected that such conversion will slacken off because of the activity of Muslim and Hindu leaders. George Sammy introduced the topic of reconversion, to which Makhan Dubé added that Hinduism has some militancy in Trinidad, which may lead to reconversion.

Many Presbyterians are taking part in Hindu ceremonies, Mr. Bhattacharya noted. People who were once converted, can always reassimilate to Hinduism by performing *kathas.* George Sammy observed that there is great religious tolerance in Trinidad, to which Mr. Bhattacharya added that in India, anyone can watch, but no one can participate of their own volition in different religious ceremonies.

George stated how degrading it once was to have an Indian first name. This is no longer the case.

Makhan Dubé regretted that the Indians are always looking toward India. Indian independence in 1947 gave people here a feeling of optimism. Then many became bold enough to press for their own organization. As long ago as 1886, the East Indian National Association of Trinidad was formed with the aim of helping the Indian Nationalist Movement in India. It had great leaders, such as David Mahabir. All fought for the removal of passes from estate Indians, badges, and other symbols of indenture. Peter Dubé added that to progress socially some people made their names less Indian—for example, Ganesh became Gunness. George Sammy admitted that there was a time when he felt ashamed that he knew and spoke Hindi. How many Indians in the Presbyterian Church have an Indian first name, he asked?

James Sammy retorted, we live in a *western* world, under the influence of Europe, the United States, and Canada. We in the West Indies are more European than Indian. It has only been since independence that India has been taking an interest in Trinidad.

Mr. Bhattacharya commented that British Guiana was the first place in which he saw the *oronhi* and dress worn by Indian women. This came about historically because the bales of cloth given to the indentured workers were not wide enough to be made into saris. The tying of the *oronhi* was learned from Portuguese women. Mr. Bhattacharya sees this dress as symptomatic of a conservative attempt to re-create clothes worn in India. Makhan Dubé added that some women wear similar clothes even in India, but Mr. Bhattacharya argued (incorrectly it subsequently appears—see glossary on *ghangri*) that Rajasthan, Delhi, and the Punjab, where women's clothes are closest to Trinidad usage, were not source areas for the indentured migrants.

Mr. Sookoo, who told us that he had sat for 17 years on a church board, stressed that Hindu high castes converted to Presbyterianism only as isolated cases.

Mr. Bhattacharya added that Roman Catholic missionaries to India had tried to convert him and his sister, knowing that, if you can first convert a Brahmin, others will follow their lead. Mr. Bhattacharya revealed that he had been criticized for holding his Gita class at Débé in the low-caste Seunarine Temple.

James Sammy described the rush to Hinduism and Islam as largely political and recent. It represented an attempt to identify the self with Indianness. Mr. Sookoo pointed to the importance in 1950 of the merging of two Hindu groups, the Sanathan Dharma Maha Sabha (SDMS) and the Sanathan Dharma Board of Control. By 1952, the SDMS was founding its first schools. The merger had been achieved through pressure brought by a group of young men in 1949; it was not the result of Indian independence or pressure from outside the society.

What is the future of the East Indian community?

Mr. Bhattacharya asked whether it was Indians in San Fernando who converted. Or was it the more likely scenario that converted people came to San Fernando? Mr. Sookoo commented that many San Fernando Christians have moved to Port of Spain because of the lure of city amenities and for their children's education.

Colin raised the issue of intermarriage, and Makhan Dubé claimed that it will occur more often whether we like it or not. The greatest tragedy for any Indian family in Trinidad is for a daughter to marry out of race; worst is a black, then white, and so on. The will to resist now lies in the older people rather than the young, George Sammy observed. Mr. Bhattacharya noted that very poor Indians are intensely conservative. Poor Indians suffered greatly at the hands of the planters over drink and women. James Sammy observed that the figures for racial intermarriage will go on increasing, despite the fact that Hindus, Muslims, and Presbyterians arrange marriages for their children.

James Sammy outlined the problem for the educated East Indian coming back from abroad to look for a wife, but Mr. Sookoo countered this by pointing out that there are so many opportunities for Indian girls now that an educated boy can find a comparable wife of his own race. James Sammy concluded that as long as a strong feeling of being Indian exists among East Indian groups, there will not be so much out-of-race marriage.

Colin asked about inter-race friendship, and James Sammy said that the predominance of one or other racial group depended on the area. In outlying areas, clubs for cricket tended to be racial, for example. Mr. Sookoo noted that doctors' or lawyers' cricket matches were always predominantly Indian. Oxford Club is 99 percent Indian. Makhan Dubé commented that Indians always prefer going to Oxford Club rather than the Palms Club, which is mostly black and is associated with the Oilfield Workers Trade Union.

Makhan Dubé observed that personal friendships were the basis for home visits, particularly in town and among educated groups. The tendency to mix socially was growing, but not for marriage. Mr. Sookoo added that there was not much visiting in rural areas; one might visit one or two special people of other races.

Makhan Dubé suggested that East Indian culture had things to offer Creoles in the shape of family life and thrift. George Sammy chipped in to remark that blacks are now saving hard to educate their children. One thing from the discussion is clear, and that is that Indian objection to marrying blacks is not because they lack a sense of family responsibility. Everyone agrees that it is pure prejudice—"beauty, hair, aroma."

How might Indian culture be incorporated?

Mr. Bhattacharya drew a distinction between a nostalgic longing for India and actually achieving specific aims. There is conflict between the uneducated guardian of Indian culture and the enlightened teacher; conflict between the emotions and practice. In Trinidad the Ramayana is studied only by Brahmins (it is the reverse in India), and nowadays read solely in Hindi. Sanskrit has been forgotten, and a vacuum is being created.

James Sammy referred to the suggestion by Eric Williams that a chair in Indian Culture should be set up at the University of the West Indies, St. Augustine, to which Mr. Bhattacharya rejoined that it will remain a suggestion and will not be revived until the next election. He continued that Indian drama is dead in Trinidad, and music three-quarters dead. Hosay,[64] once a festival of mourning, has been turned into a carnival. Trinidad Muslims do not sing, or even like music, but many Muslims are musicians in India.

Mr. Bhattacharya referred to the reactionary excitement of revival that explains the rash of Hindu temples and schools. He sees the process as a series of flashes that generate more heat than light. Hindus are trying hard to compensate for all they hadn't done before. Filling temples on Sunday mornings is a reaction. It also involves a constant comparison with the Christian Church—Mr. Bhattacharya is interested in Christianity, not the Christian Church. The massive increase in communication involves religious books in English and includes books in English on Hindu texts. Indian newspapers never came to Trinidad in the past, and people were unaware that English-language papers were read in India.

James Sammy noted that in the old days, the Presbyterian Church printed articles in Hindi. There is a tendency among the Indians for an individual to finance a temple, and not to collaborate with others, so that only the donor's name will be associated with it in future.

Should there be a Hindu theological seminary?

Mr. Bhattacharya commented that there are no theological colleges in India, no Archpundits, and no nationally instituted Pundits' Councils. Hinduism embraces atheism, Jainism, and Buddhism, so how could there be a theological college? Hinduism is a way of life; Hindus live the life. This is how it survives in India and the rural areas. A *saddhu* is an abnegated Hindu, who performs his own cremation and departs. To the world, his is a dead body.

But Mr. Bhattacharya revealed that he has been consulted on the question of a seminary. The theological college would be a dormitory where Brahmin children would live, get orthodox training, certain certificates, and robes. It would not be possible to practice without a certificate. This is a westernization of Hinduism, he says. And why, he continues, do all Hindu marriages take place on Sundays, instead of on the astrologically propitious day for the couple?

Mr. Sookoo objected to the establishment of a Hindu theological college that would perpetuate Brahminism. Mr. Bhattacharya said he had not found anyone who is diffident about caste in Trinidad, but many are arrogant. The high castes are arrogant about it, but the low castes are not so bothered. Caste is silently felt, particularly in the upper bracket. For Creoles a comparable issue is color. Mr. Bhattacharya commented that a Samaroo is a Samaroo, and there is no need to add Maharaj to emphasize his Brahminical status. He can remember two or three taking the name Maharaj in the last three years since his arrival in Trinidad from British Guiana.

Is there an elite?

Vilma Dubé observed that top people who are more westernized are more acceptable to society as a whole—Dr. Winston Mahabir, for instance. The non-westernized are accepted with reservations. Throughout the whole of Trinidad, it is acknowledged that top people have money. The old families were the Scotts, Sherlands, Georges, Maloneys, and Dopsons, and the French Creole De Verteuils. Some people have a flair for creating the aura of the old families—they are good, swift copiers of the French Creole system.

Makhan Dubé sees the Indians as looking forward, whereas Creoles are looking back, because their position is crumbling.

James Sammy noted a movement toward integration in Trinidad, in a political and social sense—an observation to which Makhan Dubé added the rider, "a certain amount of social integration despite politics." Mr. Sammy accepted the amendment.

Thursday, June 4

Colin: *Rosalind Ramsamooj on marriage*

Rosalind Ramsamooj came to see us. We asked her about marriage and caste. She says she does not want to get married yet, because she wants to do something with her life. She is determined to study hairdressing and beauty in the United Kingdom—she had previously said Canada. Rosalind claimed that she would rather please her parents than herself when selecting a marriage partner, but she would consider caste, preferring not to choose a Maharaj. To marry a Maharaj could be construed as social climbing, and they would make you feel an outsider. Rosalind, however, claims not to know her caste.

Post-focus group—Hansar and Bram

Bram and Hansar came round for the evening, and we went over the focus group at George Sammy's. Bram was disappointed with the discussion—in his view, nothing new was revealed. He felt, in particular, that James Sammy, the teacher, had been on the defensive over Presbyterianism.

We talked about the traditional East Indian women's dress: the *jhula* (bodice, with sleeves to wrists) and *ghangri* (skirt, with hem at ankles). All this was worn with a full-length petticoat, so that the skin was completely covered. Until about 10 to 15 years ago, the *dhoti* was worn a lot by men, especially for working in the fields. Now it is worn only by Indians from India, and on religious occasions. According to Bram, trash in the cane fields used to cut the legs of the cutters and loaders, and therefore trousers replaced the *dhoti* for work and soon became worn all the time.

Matti kore (when the marriage pole is planted in the *maro*) is for women only, and involves the ladies in singing and making fun. Often a hole is made in the earth, and a coin stuck in a corner of it. The bride-to-be has to search for the coin blindfolded. If she can find it quickly, it reveals her artistic ability and sensitivity. Meanwhile, the women sing and dance.

The Bombay Indians are mainly engaged in private business. Some sell jewellery from glass display cases on the High Street. There are approximately 25 families in San Fernando, perhaps more. They remain aloof from Trinidad Indians and get their brides from India.

We talked about San Fernando's top people. Hansar immediately said, "Men with money." They are respected because they are smarter than most in making money, and therefore must be smarter in reaching God. When it came to specifics, Binie Maharaj and Mr. Gopaul were mentioned—both are respected in San Fernando, but not in the countryside. If you are thinking about dutiful Hindus, then Ramnath Maharaj, Mr. Rampersad, and Bissoon Pundit would count..

Gopie is not respected among Hindus. Up to five or six years ago, he was highly respected, but his big mistake was to join the PNM. Gopie says that Indians need to be represented in the PNM in order to get a square deal, but Bram and Hansar regard him as an arch hypocrite.

Saturday, June 6

Gillian: *Syed Mohammed Hosein, Imam ASJA mosque*

We had an interview in Prince of Wales Street, just south of the Mucurapo Street Market, with Syed Mohammed Hosein,[65] whom we had met at George Sammy's focus group—though he said nothing on that occasion. During our interview with him, it emerged that he is terminally ill. Both his parents had migrated from Gonda in India. His father had had a misunderstanding with his own mother, and had quit, leaving a son behind in Gonda. Mr. Hosein visited his half-brother in 1934. Mr. Hosein's father found that he had more opportunity to make money in Trinidad than in India. Although he talked about taking Mr. Hosein to India for his education, they never went.

There were mosques in San Fernando before Mr. Hosein was born, and perhaps as many as 20 in the whole of Trinidad. For example, there were mosques in St. James, Port of Spain, in Chaguanas and on Prince Albert Street, San Fernando. In San Fernando there are now two mosques—Anjuman Sumat al Jamaat (ASJA) and the Trinidad Muslim League (TML).[66] This split represents a difference of opinion, which led to the second mosque being formed. The Muslim League Mosque is the original one in Prince Albert Street.

There are some Shia followers here, but they cannot reveal themselves. Mr. Hosein is the imam of the Mucurapo Street (ASJA) mosque (Plate 7). There are no religious images in his house. In the corner of the room, in which the interview took place, there was an earthenware pot for burning wood or incense. Mr. Hosein has a black servant. As imam, he conducts prayers and looks after the welfare of his flock.

Muslim prayers should be performed five times daily, but most people do not have time to carry out the entire sequence. Friday is the Muslim Sabbath, with prayers in the morning, at midday, in the afternoon, at sunset, and in the night. If Mr. Hosein cannot attend, someone deputizes. Most Muslims live away from the mosque, not clustered around it. The Friday Jumaa namaz (service) at 12.30 p.m. sometimes has between 200 and 300 people. Workers are given an hour or so off for worship.

Plate 7 Anjuman Sumat al Jamaat (ASJA) mosque, Mucurapo Street, San Fernando. The white building with green features is oriented toward Mecca, and out of alignment with the street

Mr. Hosein went to India on a religious mission, wanting to meet the stalwarts of Islam. A few Trinidad Muslims have made the *Haj* (pilgrimage) to Mecca, and can then be called *haji*; when they return they are highly respected in the community. But it is much easier to make the pilgrimage now, because there is more money. Nevertheless, some people managed to go to Mecca when Mr. Hosein was a child; it took six months by sea.

Orthodox Sunni Muslims do not recognize Hosay, but in the same way that Roman Catholic Carnival has been absorbed by others, so members of all communities join in the celebration of Hosay. It is an annual fête, introduced by Shias when they came from India. Now all *tadjah*–makers[67] are Sunni, and no one in Trinidad will admit to being Shia.

Eid ul Fitr is a popular Muslim festival, a time for forgiving feuds. It occurs at the end of Ramadan, and involves a service in the mosque with celebrations afterward. Bakra Eid (Eid ul Doha) occurs approximately two months later. It involves animal sacrifice and almsgiving to the poor, and commemorates Abraham's sacrifice of Isaac.[68]

Ramadan lasts for 30 days, and one should fast not only with the stomach, but also with hand, tongue, eyes, and feet, thus purifying oneself completely. Many Muslims—even little children—keep this up. Taraweeh are the prayers said at night during Ramadan, during which 20 genuflections are to be made. There is no compulsion to fast during Ramadan. You only do so if you can manage it.

The five pillars of Islam are: *kalima* (faith), *namaz* (prayers), *roza* (fasting), *haj* (pilgrimage), and *zakat* (compulsory charity). Mr. Hosein asserts that in Islam there is great respect for Christianity, but not the other way around. There has been relatively little conversion from Islam to Christianity in Trinidad. The main reasons for conversion were given by Mr. Hosein as economic betterment and marriage—that is falling in love with a Christian who insists on conversion. The latter happened to his own son, though he has, in the end, remained Muslim.

The Koran says that a Muslim may marry a Christian or a Jew, that is, someone who believes in God. But the really strict view is that non-Muslims should convert on marriage. There is no compulsion, in Trinidad, for children of mixed marriages to become Muslim (as there is among Roman Catholics), though they should receive instruction in Islam. Children of Muslim-Christian parentage tend to have a Christian first name and to go to church on Sunday.

There is no baptism for converts, but all must learn to say the following:

I believe in God
I believe in his angels
I believe in all his revealed books (Koran and Bible)
I believe in all the prophets that came into the world
I believe in the last day (judgment)
I believe that God knows all my good and evil deeds
I believe I will rise after I am dead

Children of Muslims are taught the above articles of faith at home and must recite them before going to sleep.

Mr. Hosein speaks Hindi and Urdu. Urdu was introduced to India by the Muslims, who drew from Persian and Sanskrit, as well as Hindi. But, in Trinidad, Sanskrit words in Hindi are not understood by an Urdu speaker. Much English is used in Trinidad's mosques nowadays, because children can speak little Urdu. In Suriname, every child can speak Hindi or Urdu, but it is very different here. In San Fernando Urdu is taught only in the ASJA primary school. But Mr. Hosein does not think that Urdu will die out, because of the influence of Indian films, visiting actors, and religious men. Services in the mosques are carried on in a mixture of English and Urdu.

In the past, Muslims used to marry young. Child marriages, common earlier, are no longer so. Nowadays brides are 17–18 years old. "Hindus have abused this thing for hundreds of years." No courting is allowed, following the injunction of the prophet. The Koran says that it is compulsory for a healthy man and a healthy woman to get married. It is up to the "parents to look for a decent and respectable boy or girl." Among Muslims, both sets of parents take the initiative, whereas with Hindus, usually it is the boy's parents. The dowry is optional (whereas among Hindus, girls' parents often take the initiative in order to avoid the payment). Muslim boys or girls can choose a partner and then let their parents get involved. But they cannot go on dates without a chaperone. Muslims would rather marry a Hindu than a Christian, "because all our forefathers were Hindus once."

Anyone in the Muslim community can perform the marriage ceremony; it is not necessary to involve the imam, and marriages customarily take place at the bride's home. Marriages can be performed by proxy. The name of the pair is called. The question is put, "Are you, so and so, prepared to marry so and so, and give so much dowry?" This is repeated three times. If the girl answers, "Yes," then they are married, but there must be two witnesses. Muslim marriages were legalized in 1935, Hindu marriages in 1946.

Mr. Hosein is a divorce officer, and has occupied this role since 1935. There is no need to be an imam to hold this position, since the post is filled by the government. Fewer Muslims than Christians are divorced, though it is often said that Creoles become Muslims because of the ease of divorce. In reality, divorced Muslims are practically ostracized from Muslim society, and it is difficult for a divorced person to get another spouse. Divorce is pronounced in the same way as marriage: "I have divorced you," is repeated three times; it may be repeated once a month, in which case the divorce becomes valid in three months.

There are no grounds for divorce among Muslims. If you find your wife committing adultery, you should forgive her. But, if you cannot forgive her, then you have every right to divorce. Nowadays, all Muslim marriages are registered, but the divorce law (introduced in 1935) still applies. For the divorce law to apply to mixed marriages, the ceremony must have been conducted according to Muslim rites. Conversely, if two Muslims marry in the Warden's Office in a civil ceremony, they can be divorced only in the law courts. The most common ground for divorce among Muslims is incompatibility. There has been a great decrease in divorce in recent years, largely because improved educational standards breed a higher regard for marriage, but also possibly because young people are now choosing their own spouse. However, Mr. Hosein thinks that parents will go on having a strong influence on the choice of marriage partner.

Muslims consider themselves part and parcel of the Indian community. At any function, both Hindus and Muslims are invited. Muslims must have meat—chicken, goat or sheep, but pork is prohibited as are alcohol and all games of chance. Most of the San Fernando soft drink manufacturers are Muslim—for example, Mr. Sheik Rahaman, the nephew of Mr. Hosein and son-in-law of Mr. Jaleel[69] (the original owner of the bottling plant).

In Mr. Hosein's opinion, the Muslim faith is gaining strength, because it is being preached to everyone. About 50 to 60 Creoles in Trinidad have become Muslim; 6 to 8 in San Fernando. Muriel Donawa, the politician, has been a Muslim for about 20 years. It is hoped to bring some preachers from Algeria and other countries where the "black man rules" to appeal to Trinidad Creoles.

Colin

Mr. Hosein served on the East Indian Advisory Board in about 1933–34, having been nominated by Governor Fletcher. In 1934 and 1936, Mr. Hosein went to India by boat and traveled extensively by train. In 1935, he brought back the first Indian film[70] to be shown in Trinidad.

Monday, June 8

Gillian: *Muslim wedding in Princes Town*

On the evening of Saturday June 6, we went to Princes Town with Peter Dubé to see some of the preparations for a Muslim wedding to be held the next day. The Dubé family goes back three generations in Princes Town. Two sons, Makhan and John, remain there. Both are esteemed members of the community, and are referred to as "Doc" and "Head Teacher," respectively. On Sunday Peter was going to take us to witness the marriage of the daughter of a childhood friend, a Muslim goldsmith called Tam. His family was one of the first in the street to have a bathtub. The Dubé boys used to go there to wash as children. Vilma Dubé, describes the family as "westernized Muslim."

The wedding was between Tam's second daughter, aged about 21, and the son of an imam from Iere Village, on the edge of the town. The couple had met and wanted to marry, and the union was approved by both sets of parents. The bride is a teacher and the young man a mechanic. However, despite the approval of both families, there has been considerable tension and conflict. Tam's elder daughter is not yet married and is very westernized. She is unhappy about the match, because the groom is staunch Muslim and rather old-fashioned. Moreover, she feels uncomfortable that her younger sister is marrying before her. In addition, there has been an element of east-west conflict over the marriage ceremony, with Tam's family wanting a westernized wedding, with music and a wedding cake, and the groom's family preferring a more strictly Muslim ceremony.

In the event, a compromise was reached. On the eve of the wedding, the groom's family held prayers at home. At the bride's home, villagers gathered to prepare the wedding food, but, in accordance with the imam's wishes, there was no music to accompany them through the night. The parents had originally agreed that the wedding ceremony would take place inside the house, though Tam wanted it outside where all could see. On the wedding day, the imam (the father of the groom) suddenly announced that he wanted a double wedding—that is, a wedding by proxy in the traditional Muslim manner. In such a case, the boy would remain outside the house, and the girl inside, and they would then be married through a messenger. There was much argument between Tam and the imam, but Tam won this time, and it was agreed that the couple should be married on a settee in the front room.

As we arrived on the Saturday evening, villagers were gathering at the bride's home. Some helped prepare food, others tried to prevent leaks developing in the bamboo and canvas tent, in which chairs had already been set out. The main room in the house was elaborately decorated with sprays of fern and artificial vanda orchids on the walls. At one end stood a table displaying the wedding presents—vases, a dinner set, glass dishes, cutlery. In pride of place stood the wedding cake—three huge tiers set on a glass tray complete with plastic swans. A life-sized bunch of roses had been worked into the icing on one side of the bottom layer. The pièce de résistance, on the top tier, was a plastic archway with a distinctly western bride and groom in the middle.

The main preparations took place under the house. There, men of all shapes and sizes were plucking, singeing, and chopping up dozens of pink chickens. Although a few women were helping or advising, this was largely the men's affair. "Head Teacher" had the unenviable task of gutting the fowls. We looked on in amazement. Suddenly we were able to appreciate the value of communal effort at such a time. The bride's parents were paralyzed into inactivity, while good friends ran the show.

In the area under the house, too, stood the vast containers for boiling the rice and chicken and for mixing the flaky *parata roti*. *Roti*-making for weddings is almost a profession. It would seem that no chicken-chopper would undertake to prepare *roti*. Instead, a serious, muscular young man arrived, especially for this purpose. The mixing-pot was set in a used car tire, to prevent it from tipping. Then, pound upon pound of flour and water were added to the pot and kneaded by hand into soft dough. In all, 70 pounds of flour were used on this occasion. The preparation of the dough was arduous; it was tiring just watching the *roti*-maker, elbow-deep in dough, blending and mixing it all by hand. Sweat poured from his face and arms, and, like a professional wrestler, he had to retire at intervals to be dried off by a friend.

When this stage was over, *ghee* was poured into the pot, in order to drag the dough away from the sides. Then long tables were set out and covered with a clean cloth made from flour bags. While the *roti*-maker took out two-pound chunks of dough, his associate formed a large ball with it, and set it out on the table. Having done this, they then took each ball, flattened it, smeared it with *ghee*, and slitting from the edge to the center, rolled it into a cone, which they left to stand.

By this time, a small pit had been dug in the ground, and wood had been laid inside it for a fire. Over this was placed a three-foot wide iron griddle or *tawa*, which was cleaned and smeared with oil. Once the cones of *roti* had settled down into rounds, and the griddle was hot enough, the roti-maker rolled out the dough again until it was about 12 inches in diameter. Then he lifted it up and stretched it until it formed a circular sheet, which he threw down on the *tawa*. The other man greased and turned it, until it was cooked and somewhat shredded. It was finally heaped onto a wooden tray, where it would stay warm until the next day. We left for San Fernando at about 2 a.m., by which time the preparations were well under way.

The wedding on the Sunday was a modest affair. There were many Creole participants, as there had been Creole helpers the day before. Only one guest wore a sari. The bride was dressed in an elaborate white, western-style wedding dress with a veil, while the groom wore a dark suit, white shoes, and a white turban.

When the groom's procession of cars arrived, they stopped outside the bride's home. The formal meeting of the fathers, using the crossed-hands Muslim greeting, then took place. The lineup of friends (all male) behind the bride's father was reminiscent of the Hindu milap. After entering the house, the groom sat on the settee in a hot crowded room. Then the bride entered, and sat on his left. The marriage ceremony was conducted in Urdu, all very short and simple. A few questions and prayers were followed by an explanation given in English. Finally, the bride and groom ate some sweetmeats and drank a soft drink from the same glass, the husband first.

When the ceremony was over, the couple remained seated while friends came up and spoke to them. The groom gave his wife a wedding ring. Soft drinks, cake, and Indian sweets were handed around. Sometime later, the couple left for the groom's home, accompanied by many others (Plate 8). They would be formally received by the groom's parents, but were not going to live there. They have a new house in Diamond Vale, and will probably honeymoon in Mayaro.

Plate 8 Muslim bride and groom leaving the bride's home after their wedding, Princes Town

Vilma Dubé

The remainder of our visit to Princes Town was devoted to a conversation with Vilma Dubé. She deplores the typical Trinidad Indian attitude to India. When India plays the West Indies at cricket, the East Indians en masse cheer for India. She does not approve.

Vilma stresses that she doesn't reject her Indian heritage. She feels some attachment to India, and an interest in its history and thought. But she regards Trinidad as part of the western world, above all. When she was young, her family lived in Penal, but they bought a house in San Fernando so that the children could attend school during the week. A white Barbadian nurse was installed to look after them. Vilma won an Indian government scholarship to an Indian University, but on her return, the Indian High Commissioner was offended because she did not wear a sari. Vilma said that she went to India to be educated, not to be converted to an eastern way of life.

Although of Hindu origin, Vilma is now Anglican, having been educated in Presbyterian and Roman Catholic schools. Her brother-in-law, Peter, is also an Anglican, but married his wife according to Muslim rites, and he is now interested in Hinduism. His wife is an active member of the Presbyterian Church, though she has never been baptized as a Christian.

Thursday, June 11

Gillian: *Chaguanas and Felicity*

We went to Chaguanas with Bram (Plate 9) and saw Lion House, the home of V. S. Naipaul's maternal grandparents, before traveling deeper into the Caroni cane lands to Felicity (Klass's Amity). Felicity turned out to be a prosperous village, set in flat land, with cane on one side of the road and rice on the other. Most of the houses were made of concrete; only a minority were of wood. "Jangli Tola" looked very poor, but occupied only a smallish area.[71] The best houses were located on the main road, and there were some blacks standing around at the Junction. Many wattle, mud, and thatch *ajoupas* were in lovely condition, even painted. We located the temple, Maha Sabha School and infant school, where children were sleeping on tables, benches, and on coconut mats on the floor. Fringed by the lagoon and the river, the countryside adjoining the village was flat and wild. Bullocks were pulling carts, a man was fishing in the river, and there were rice fields beyond.

Friday, June 12

Gillian: *Temple in the sea at Waterloo, and Gran Couva*

We went with Bram and his brother, Hari, to see the temple in the sea at Waterloo on the Gulf of Paria. Built single-handedly by an

Plate 9 Bramadath Maharaj near Felicity

old *saddhu*, its base was constructed of kerosene tins filled with concrete, rubble, and pitch. The temple erected on this platform is splitting across the walls and floor, but the old man keeps repairing and enlarging it. At the entrance, there is a tiny *kutia* (shrine) for prayer. Inside the *ashram*, there is a room containing a bed and another altar. A service is held there once a month at full moon. This is a private temple. The *saddhu* has built a causeway of bricks, stones, and broken bottles, but when the tide is in, it covers the path. His construction gets swept away at a fast rate. As we left, he was picking up stones, putting them into his bicycle basket, wheeling it along the path and throwing them down at a weak spot. In this painstaking way he has constructed the whole edifice.

Hari drove us inland to the coffee- and cocoa-growing area around Gran Couva—so different from the coastal area, hilly, wooded, luxuriant, quiet, and beautiful. The cocoa pods are big and maroon-colored, and the tiny green coffee beans grow in clusters along the stems of the plants. We saw two cashew trees with thin golden and red fruits and green nuts hanging below, and mile upon mile of hibiscus hedge.

Saturday, June 13

Colin: *Gobardhan Jerrybandan on religion and caste*

I have had a visit from Gobardhan Jerrybhandan, who lives on Cooper Street, between Coffee Street and Rushworth Street. He had read Herman Roop Dass's article about us in the *Trinidad Guardian* ("East or West Indians?" Wednesday, June 3, 1964) and wanted to give us some information. We had been concerned about the publicity, and here was our first caller. He tells me that Hindus and Muslims don't trust one another. He eats neither pork nor beef, and he invited me to inspect the superior quality of his teeth (which are mostly gold-filled, anyway). On Hinduism, he says, "It is a way of life; because of the culture." He got married two months ago to another Maharaj: "Maharaj should marry Maharaj." Purebred horses were referred to. "The caste system is not the best thing, but I'm in it, so what can I do?"

Gobardhan (he explains that *gobar* is the dung that disinfects the smooth mud floors that East Indians lay out under their houses) had never seen his bride before he was taken to meet her by his father. Gobardhan thought she was nice. The parents had approached one another to make the match. The couple's horoscopes were examined to test their compatibility. Once the engagement was celebrated, the couple could go to the cinema together, and speak in private at home. But if the girl had not been a virgin on marriage, she would have been sent back to her parents.

Five brothers live in the family home, of whom four are married. So the household unit has 30 people living there altogether.[72] His wife covers her face as she approaches her father-in-law. She will not speak at home in the presence of her older brothers-in-law (the *barka-*

chotki avoidance relationship). Menstruating women may not go into the *puja* room or prepare *persad.*

Hindus who eat pork but not beef are "Chamars." He would not trust his wife with a Chamar: "Quicker trust a Maharaj than a Chamar. Chamars think low and act low." "Quicker trust a parrot than a Muslim." Gobardhan claimed that Chamars are lower than Muslims. A Chamar who does not eat pork can improve his caste. Singhs acting as pundits become Brahmin. There is no untouchability in Trinidad. If Maharaj or Chattri do low things, they are considered no higher than Chamar, for example a pork-Maharaj. "Caste is breeding. A Maharaj feels funny about cleanliness in the home of a Chamar."

Sunday, June 14

Gillian: *Mr. Jogiesingh, California*

We were invited to visit Mr. Jogiesingh, who had also read about us in the *Trinidad Guardian.* He was living in a poor sugar-welfare house at Dow Village Settlement, California. The concrete house had been built with a government loan, which Mr. Jogiesingh must pay back in 20 years. It was completely unpainted, the bare walls cracked and scribbled on. There was no banister on the stairs leading up to the main rooms on the first floor, which made them dangerous for his young children.

Like Jogiesingh, his wife is a cane worker. They have six young children, and the smallest at nearly two is still being breast-fed. The household manifested extreme poverty—there were hardly any chairs. Mr. Jogiesingh has a hobby—collecting Amerindian artifacts from the countryside. It was about these that he had wanted to speak to us.

Wednesday, June 17

Colin: *Recollections and reflections*

Indians are impressed by educational qualifications, yet they have no idea about Dr. Rudranath Capildeo's specialization in mathematics. Anything that is abstruse is fine. Dr. Capildeo refused to read the party manifesto in San Fernando, because he said the people were not educated enough to understand. This was accepted. Capildeo is basically an authoritarian, charismatic leader.

The cohesion of the Indian community is racial; it is based on form, not function. The only function is politics, and then the Muslims are largely excluded from the DLP (though they are culturally more "Indian" than the Presbyterians). It is said that the Muslims are in two political camps, represented by Saied Mohammed[73] (PNM) and Tajmool Hosein[74] (DLP). Only the Hindus are staunch "anti-Creoles," politically and culturally.

Authoritarianism, tempered by an imposed "democracy and the rule of law," has led to factionalism in Indian institutions, such as the Maha Sabha, Arya Samaj, the Cane Farmers' Association, the Muslims, the temples in San Fernando, and the DLP.

The materialism of the west and spiritualism of the east were thumped home by Tajmool Hosein and Haji Francis last Sunday (June 14). Yet many Indians are incredibly money-minded—Bhadase Maraj (president of the Maha Sabha), Binie Maharaj (vice president), and Mr. Gopaul (vice president). Bhadase's house has its own *mandir*, with slot machines in the undercroft; Mr. Binie combines the temple with horses. There is juxtaposition of crass materialism and spurious spiritualism here, with spiritualism often an excuse for meanness.

Trinidad's Muslims were given BWI$50,000 (£10,000) by the government toward the building of ASJA College in San Fernando. It was essentially a political payment for electoral services rendered. Bhadase donated several thousands of dollars toward the Tunapuna mosque; was this a political ploy as well?

Mrs. Kelkar is going to give a Hindi and Hinduism class in the San Fernando *ashram* on Fridays for boys released from ASJA College (when the Muslims go to the mosque).

San Fernando has no Maha Sabha branch, though there was a San Fernando Hindu Sabha in the 1920s.[75]

Rich Indians are respected; but Hansar is not considered because he is poor—women give him a few shirts, because Bram urges them to do so.

In an Indian-language film I saw with Bram, some girls were teasing the heroine about her lover. Translating for me from Hindi, he whispered, "They giving she a little picong."[76]

Thursday, June 18

Colin: *Bram and Hansar*

We had a discussion with Bram and Hansar about a variety of aspects of Hindu life. The word "shipmates" came up. The term used for those who traveled from India together is *jahaji bhai* (shipmates) from the Hindi for ship, *jahaaj*. Bram's father's shipmate lived in Hermitage, and was treated by Bram as though he was his father's brother (*dada*).

Muluk is the Hindi word used to refer to India. Therefore, someone from India is *mulki* or *mulkin*. Hindus may not marry any member of their family or adoptive family. But a brother can, and frequently does, take his dead brother's wife, in order to keep her and her children within the family.

Kathas, or rituals at which Brahmins are fed, are held quite regularly in San Fernando. Normally it is a small affair. Mrs. Rampersad often invites Bram to eat a little *persad* to mark the occasion. Brahmins are also fed to honor the dead; ten days after the death of a woman, but thirteen days after that of a man. Bram is not too keen on these occasions, preferring to sit among his own friends "and make a bit of fun," rather than joining the selected Brahmins.

There are four pundits living in or near San Fernando: Ramnarine on Hickling Street; Bissoon in Gooding Village; Bisnath on Farrell St; and Moon or Mohun at La Romaine. Jankieprasad Sharma, who lives in Débé, is the Dharmacharya (spiritual leader) of Trinidad. He officiates at ceremonies for Binie Maharaj. Bram and his family have a pundit from Penal. Both bride and groom are usually baptized shortly before marriage so that the pundit who officiates at their wedding will be godfather to both.

Dharam Pundit from Diamond Village had his children take the sacred thread (*janeo*) about two years ago. Bram is considering getting a pundit to be his godfather. But he will not take the thread, because it is too difficult to live up to—one can't, for example, eat meat. Ramnath Maharaj has taken the thread. He fasts one day each week, abstains from alcohol and meat, but doesn't practice as a pundit. He prefers to observe all the religious festivals in a private way. The Seunarine sect has its own pundits, but they are not Brahmins. Nearly all Seunarines are drawn from the two lower *varnas* (Vaishya and Sudra).

Talk then turned to place names used by Hindus, though they are rapidly falling out of use. Pity Book (*petit bourg*, little town) was used for San Fernando; *Sahar* (the Hindi for town) meant Port of Spain; *Nesan* (a corruption of Mission[77]) was Princes Town. All the villages have alternative names as well: Chauhaan = Chaguanas; Sally Grand = Sangre Grande; Tunpoon = Tunapuna; Pinjal = Penal; Pont a pew (Pointe-à-Pierre).

Susamachar, the name of the Presbyterian Church in San Fernando, is the "Church of Good News or Tidings." The catechist was paid a certain amount per head for converts.

Hansar taught himself Hindi without the help of a pundit. He approached several, but they brushed him off. He would read a bit of the Ramayana and the Gita. He was about 15 to 17 when he began to take an interest in Hinduism, doing a bit of meditation, but all on his own. All his ideas were written on scraps of paper; he now uses the same method for preparing his lectures. Hansar was not particularly good at school. He was afraid of the teacher and had no interest in study or the search for knowledge. However, he was interested in the stories his mother told in Hindi.

While we were talking about mixed marriages, Hansar commented that it was better that a couple should forget about religion if this would maintain their happiness; there is no point in being stubborn and bigoted about religious differences. Bram stressed that in racially mixed marriages, one race will always try to express its superiority over the other. He had no sympathy for our suggestion that if two people were well matched, race is no obstacle.

After I had written up this discussion, I added the following reflections. In Trinidad's plural society, one must take care to distinguish between politics and power. The East Indians seek power within their own community; for example, Brahmins among the other castes. Power and wealth are important, but not power and ideas—ideas for Indians are religious and philosophical, but there is little interest in aesthetics.

Racial feeling is strongest among Indians. Hari Maharaj commented, "everywhere there are blacks in the world there is trouble," and it is likely that this is the opinion held by most Indians in Trinidad. Blacks display a contemptuous disregard for the Indians. There is no need for open hostility, since they are in the majority. It is easy to disregard a watertight minority that accepts no intruders and lacks political leadership. It is very easy to disregard a group led by an absentee political leader, such as Dr. Capildeo, and by unscrupulous religious leaders like the headman, Bhadase Maraj.

Among East Indians, there is a fundamental misunderstanding about the working of democracy. Democracy is not conceived of as a continuous process, but as a method of beating each other at the polls, and then forming a dictatorship. Furthermore, victory at the polls is secured by enlisting racial support—so there is really no democratic process in parliament, nor is there a victory of one set of ideas over another at the polls. This is the attitude of the Creoles, too.

Saturday, June 20

Colin: *Bisram Bissoon, Corinth*

Today we went to Corinth to visit Bisram Bissoon, one of Hansar's helpers in SSDM. Bisram's father, Bissoon Lal, is a drainage worker, and his mother cuts cane. All of them work for Usine Ste Madeleine. Bisram started hauling cane with his own ox cart when he was 15. This has now stopped: he works in the factory yard in-crop, and as a planting-gang supervisor out-of-crop.

Bisram has one bull and four cows, all Holsteins. They produce 20 pints of milk a day, which he sells to customers on Coffee and Cipero Streets in San Fernando. He also owns four plots of land: one purchased from Gopaul at Marabella, and three in Corinth. Bisram keeps the cows until they die, though some people sell them to the slaughterhouse. He makes approximately $BWI 35 (£7) per week.

Bisam's parents have eight children living at home, five boys and three girls. The two oldest girls are married and live in Corinth and Tunapuna, respectively. One girl stays at home to look after the babies, while another, aged 15 or so, teaches in a commercial school in San Fernando. Bisram's parents' house is good by Corinth standards, and has many pictures of Hindu gods, such as Hanuman, and Shiva and Parvati. We had a meal of fish (salmon), and the milk was served in rum bottles.

We walked around Corinth with Bisram, and found the remains of a ten-room barracks (of the kind intended for indentured laborers).[78] Only the first and last rooms are still standing. Sugarcane grows in between, where there were originally floors. We met an old Indian, who is said to be 93, doing *jharay* (sweeping away sickness) over a small baby brought to him by a young woman. He said that he had prayed many times over Bisram's little brother. The old man referred to San Fernando as "Pity Book"; he still sells goods in the Princes Town and San Fernando areas.

Bisram owns three house lots in Corinth for his younger brothers and himself. This is an excellent example of Indian ideals and family economics.

George and Myrtle Sammy on the Madrasis

In the evening we had a talk with George and Myrtle Sammy about the Madrasi population. Madrasis are found in small pockets, for example, in Pasea Village near Tunapuna in northern Trinidad, and around San Fernando, at Ben Lomond and Duncan Village. The latter was named after Squire Duncan, a bookkeeper on the Lamont estate at Palmiste on the southern edge of San Fernando. In the 1920s and 1930s, about eight out of the twenty families in Duncan Village were Madrasi, and George's parents—both Indian-born—were the third family to settle there. The first residents were the Applewhites, who were black. The village leaders in those days were a Maharaj and two Singhs. They controlled the only *panchayat* to be held in Duncan Village in George's lifetime. This was called because of a girl's behavior.

According to George, the Madrasis came from the coastal areas close to Madras. They sold fish, drank rum, ate chilli, and chewed *pan*. These Tamil speakers were Shiva worshippers. George spoke some Tamil, but never took it seriously; he also spoke a little Hindi. He was taught Tamil on a tray of sand. There are two major divisions among the Madrasis (though they are not recognized by the North Indian castes): the Moon Sammy, who are Brahmin and the Mootoo, who are low caste. Madrasis have their own Brahmin pundits; but Chamar pundits are called *mahants*.

Myrtle's family came from Calcutta; her father was brought as a baby. He became the headmaster of the Presbyterian school in Débé. Myrtle's mother's father, too, was a head teacher. When George's father first met Myrtle, he asked, "Kalkatya?" ("Are you from Calcutta?") And when she answered "Yes," he had nothing more to do with her. At first, Myrtle's mother, known to all as Myrtle Ma, objected to George as well, saying, "Madrasi and nigger is de same ting."

Madrasis had big fights. On Saturdays they sold provisions in the San Fernando market, and then drank in the rum shops on the way back to Duncan Village. By nighttime, fights would break out. On Sunday mornings, a goat would be bought and cooked, leading to another cycle of drinking and fighting. By Monday morning, all would be peaceful and the fighters would be friends again. Woven into this cycle of community violence was the tension between two of the Madrasi men who were sharing one wife.

The conversation moved on to the Madrasi fire-pass (George had never heard of sword-ladders, which are climbed in Mauritius). Ritual crossing of burning embers with bare feet used to take place in St. James, Port of Spain. There are many Madrasis living there, though their names have been heavily anglicized. In the south, the fire-pass used to be held at Hermitage-Woodland, to the north of Débé. George's father got burnt attempting the fire-pass at Coolie Wood.

Duncan Village was very small, and the nearest substantial rural settlement was Débé, where, in the 1920s, only the postman, the stationmaster, the shoemaker, and blacksmith (and their families) were black. In the intensely Indian environment in Duncan Village, where all produce was shared, George and the black Applewhite children got on well, but their mothers fought tooth and nail. The Christian Indian catechist did *jharay* over the children. Now the village is urbanized and has become part of San Fernando. Such has been the rapidity of social and economic change in Trinidad since the 1930s.

Frank Clegorn's top people

When Frank Cleghorn gave us his list of San Fernando's top people, he named 42. Of those, 33 percent were Creole, and 66 percent, East Indian. Half those in his Creole category were white. Evidence suggests that East Indians were pressing hard for elite status in the years leading up to independence.[79]

Saturday, June 27

Gillian: *Hari Lal on marriage*

Hari Lal, who cleans the floor for us once a week, reckons that all Hindus know their own nation (caste). Hari was told his by his parents, and he will do the same for his daughter. Hari eats pork, not beef. He thinks you should marry within your own nation, as he has done.

Hari has been married for two years, to the daughter of a woodcutter. He was approached by the girl's father; they discussed the possibility of marriage; then a date was set for Hari to see the girl. He liked the look of her and agreed to marry her. She left him one month after their marriage and returned home to her parents. She did not like Hari, and was ashamed of his poor home and the fact that he had no work. Hari later learned that he was the fifth young man approached by his wife's father. She had turned down four already, and didn't like the look of Hari—but she didn't say a definite "No." After much persuasion from her father, she had agreed to marry Hari.

They were given a big wedding, considering how poor both families were. Hari's father-in-law reckons that it cost him BWI$500 in all, which is a huge amount out of a poor man's savings. It is significant that the man's second daughter was married recently to a young man with whom she was in love, but they had no large Hindu wedding. Instead, the pundit was called to the home, a small *puja* was held, and the pundit signed the marriage register. This form of marriage, known as the "table wedding," is popular among those who want to be married by religious rites and cannot afford a full-scale ceremony.

Hari regards himself as having "chosen" his bride. Our way of choosing a bride is, in his terminology, "being in love with the bride." Both Hari and his bride were disappointed with each other at first—she for reasons stated above, and he because she would not settle down to making a home—cooking, sewing, and gardening. I gather that they quarrel a lot, and that she has left him several times. She always goes back to her parents. He reckons that the modern Hindu method of "choosing" a bride is much better than the method of 50 to 60 years ago, when bride and groom saw each other for the first time at the wedding.

Saturday, June 27

Gillian: *Survey anecdote*

Our survey of samples of residents in San Fernando (about 800) and Débé (about 100), which will be carried out by assistants trained by us, has begun and will last until we leave Trinidad in early September. It takes much of our time to administer the survey and check the questionnaires for accuracy and completeness. I shall take on more and more of this responsibility as Colin concentrates on the semi-structured interviews scheduled for July and August.

Mr. Sukal told us about a questionnaire interview he gave on our behalf to a Muslim woman in Débé. The question was, "Do you think that Grenada should be joined to Trinidad and Tobago?", and he explained that Grenada was a small island near Trinidad, like Tobago. She asked, "Dey is Creoles?" Mr. Sukal said, "Yes," and the woman promptly answered, "No."

Monday, June 29

Gillian: *Frank Cleghorn*

Frank Cleghorn told us that many Trinidad Indians are disillusioned by their visits to India. Harry Pooran, for example, was snubbed. One of the people, with whom he was to stay, refused to have him in the house because of his low caste.

Frank said that he would be happy for his daughter or son to marry a black, as long as they were happy together. His grandmother, who came from India, married a black, hence the name Cleghorn.

Colin: *Mr. Cleghorn Sr. on indentured laborers*

Frank took me to meet his uncle, Mr. Cleghorn of Usine Ste Madeleine, who is almost blue-black. He went to work for the sugar factory in 1906 at the age of 13, when the Lubbock family were owners. First he was employed in the accounts department, and then as an overseer. He vividly remembers going down to San Fernando station and collecting the newly arrived indentured laborers from the railway cattle trucks. They had lists of the arrivals, giving demographic details, including caste.

Mr. Cleghorn recalled the hardships suffered by the immigrants, the overcrowded barracks where they were housed, and the brutality, especially during the crop season. He and his assistants used to instruct the newcomers how to work, for example, by grabbing a bundle of cut sugarcane and then running up the ramp and heaving it into the cart for transport to the sugar factory. But, as he commented, there was a huge difference between the demonstration and its endless repetition during a working day.

One of Mr. Cleghorn's jobs was to visit the new arrivals every day during their first year of indenture. They were given food, and a record was made of any births. From among the indentured laborers he dealt with he remembers a surveyor, who had been thrown out of caste by his family back in India, and a dentist, who had inadvertently fallen into the recruiting agent's trap. The Immigration Department controlled the indentured Indians, and Mr. Cleghorn recalled Commander Coombs, Mr. Gibbon, and Mr. Murray, who were then officials.[80]

As an estate employee, Mr. Cleghorn was well aware of the significance of differences among the Indians. Caste was always to be borne in mind, with the Maharaj caste selected as drivers, and Chamars isolated so that the other castes could not take advantage of them. Muslims were always a small proportion in any shipment, because they were said to be troublemakers. The Canadian Mission made a substantial effort to alleviate the position of the East Indians, for example, by paying to transport the children to and from school.

Notes

1. Bisram Gopie, Member of the Order of the British Empire (1962), born Ne Plus Ultra village, 1910. Accountant, involved in Indian welfare ("Who's Who," Murli Kirpalani *et al.*, (eds.), *Indian Centenary*

Review: One Hundred Years of Progress, 1845–1945, 1945, 135). Principal Officer, Sugar Industry Labour Welfare Committee. General Secretary, Sanathan Dharma Maha Sabha; General Secretary, All Trinidad Sugar Estates and Factories Workers Trade Union. Manager, Gandhi-Tagore College; President, National Council of Indian Music and Drama of Trinidad and Tobago (Carlton Comma, ed., *Who's Who in Trinidad and Tobago 1966,* 22).

2. The *oronhi* is a white veil worn by adult East Indian women, Hindu, Muslim, and Christian.
3. Mr. B. Bhattacharya, born Benares (Kasi, Varanasi), 1909. Educated Allahabad University (BA English, 1933, MA English, 1936, BEd. 1937), teacher, poet, composer, and author of books on Hinduism. Principal of several secondary schools in India, and identified as potential founder of schools overseas. Founded Tagore Memorial High School, British Guiana, 1955–61. Between 1962 and 1969 he established three colleges in Trinidad. By the late 1960s, he was a friend and confidant of Dr. Eric Williams, prime minister of Trinidad and Tobago. Celebrated his 100th birthday in India, July, 2009. (Theron Boodan, *Newsday,* June 30, 2009).
4. The *janeo* or Hindu sacred thread is traditionally taken by the three (upper) twice-born *varnas,* Brahmin, Chattri (Kshatriya), and Vesh (Vaishya). In Trinidad Brahmins were more likely than others to have been initiated into Hindu adulthood.
5. Our questionnaire surveys showed that urban and rural Hindus steered clear of Carnival, while Christian East Indians from San Fernando participated, both as spectators and as members of costume bands (Colin Clarke, *East Indians in a West Indian Town: San Fernando, Trinidad, 1930–70,* 1986, 107).
6. "Massa Day Done" was a speech given by Dr. Eric Williams at the so-called University of Woodford Square, Port of Spain, March 22, 1961, PNM Publishing Company, 1961.
7. George Moon Sammy, born Duncan Village, 1922. Educated Canaan Canadian Mission Indian School; passed university matriculation examination, 1953; Sir John Cass College, London University, BSc. Chemistry, 1957; PhD. University of the West Indies, 1966; MSc. Food Science and Technology, University of Massachusetts, 1967. Office boy and laboratory technician, Trinidad Leaseholds prior to 1953; Chemist, Texaco Refinery Laboratory, 1957–59; Research Chemist 1959–64; Lecturer in Chemical Engineering, University of the West Indies, St. Augustine, 1964–77; Professor of Food Technology, 1977–86. Senator, United Labour Front, 1976–77. Chaconia Medal, Gold, 1988 (posthumously). Died 1986.
8. V. S. Naipaul, *A House for Mr Biswas,* 1961.
9. Ganja was brought into Trinidad from India by the indentured laborers.

10. Bhadase Maraj, born Caroni, 1919; proprietor, trade unionist, and legislator. President of the Sanathan Dharma Maha Sabha, 1952–, President-General, All Trinidad Sugar Estates and Factories Workers' Trade Union 1957–; elected independent Member of the Legislative Council, 1950, Leader of the PDP, 1952–58; Leader of the DLP, 1958–59; Leader of the Opposition in Parliament 1956–59 (Carlton Comma (ed.), *op. cit.*, 1966, 165; Yogendra Malik, *East Indians in Trinidad: A Study in Minority Politics,* 1971, 84–5). By 1964 Bhadase had been ill for five years, and his power was waning. He rekindled his political career in 1967, when, as an independent, he took the seat vacated by Rudranath Capildeo (DLP), and in the early 1970s formed the Democratic Liberation Party with Lionel Frank Seukeran and Stephen Maharaj. They were totally defeated in the 1971 general elections, and Bhadase died later that year.
11. Rupert Carlyle Archbald, QC. Admitted to the Bar in Trinidad and Tobago, 1922. Chambers in San Fernando. Trinity Cross, 1972.
12. Harrichand Pooran, born Reform, 1914. Proprietor and merchant, financier, and manager of business houses: has worked for charitable causes ("Who's Who," Murli Kirpalani *et al.* (eds.), *op. cit.*, 1945, 155).
13. Simboonath Capildeo, born 1914 at Chaguanas. Solicitor and conveyancer ("Who's Who," Murli J. Kirpalani *et al.* (eds.), *op. cit.*, 1945, 137). Simboonath was the elder brother of the DLP Leader, Rudranath Capildeo, and himself the Member of the Legislative Council for Caroni South 1956–61 and Member of the House of Representatives for Couva 1961–65 (Carlton Comma (ed.), *op. cit.*, 1966, 67). He was briefly Leader of the Opposition in 1965, before resigning from the DLP. He continued in politics, switching parties, until the demise of the Democratic Labour Party at the 1976 elections. Chaconia Medal, Gold, 1989.
14. Morton Klass, *East Indians in Trinidad: A Study in Persistence,* 1961.
15. For example, V. S. Naipaul, *The Middle Passage: Impressions of Five Societies—British, French and Dutch—in the West Indies and South America,* 1962. Chapter 2 on Trinidad is notoriously jaundiced.
16. Legal registration of Hindu marriages was introduced only in 1946, a decade after the Muslim ordinance. Ten percent of Muslim heads of household and 21 percent of Hindu heads in San Fernando were in religious but illegal unions, as were 38 percent of Débé heads (Colin Clarke, *op. cit.*, 1986, 120).
17. In later life Hansar Ramsamooj (1921–2002) published *A Collection of Poems*, 1985. Hansar devoted 20 years of service to the Todd Street temple, now called the Krishna Mandir, in San Fernando, and became a licensed pundit and marriage officer.
18. The *Tulsi* tree (a sacred plant, identified with Vishnu—in Trinidad it is always referred to as a tree) is honored daily, together with the sun.

Each morning in Hindu households, members "throw *jal*" (pour holy water) from a *lotah*, while reciting an appropriate *mantra*.

19. The Ramayana, the most popular Hindu text in Trinidad, recounts the marriage of Rama and Sita, their adventures culminating in Sita's kidnapping by the King of Ceylon, and her rescue by Rama and his brother Lutchman with the aid of the warrior monkey, Hanuman.
20. Lionel Frank Seukeran, born Tableland, 1912. Educated Naparima Training College, San Fernando. Contractor and proprietor of a business house in San Fernando; prominent in various religious organizations and in public welfare ("Who's Who," Murli J. Kirpalani *et al.* (eds.), *op. cit.*, 1945, 163). Elected independent member of Legislative Council for Naparima 1956–61; DLP Member of the House of Representatives for Naparima, 1961–66 (Carlton Comma (ed.), *op. cit.*, 1966, 221). Unseated as an Independent by the DLP in Naparima in 1966, Seukeran formed the Democratic Liberation Party with Stephen Maharaj and Bhadase Sagan Maraj in 1971, only to be totally defeated in the elections that year. Chaconia Medal, Gold, 1985. An account of Seukeran's life has appeared since he died in the mid-1990s, based on his own text: Ken Ramchand (ed.), *Mr Speaker, Sir: An Autobiography of Lionel Frank Seukeran*, 2006.
21. Vilma Dubé, born Penal. Indian government scholarship (BA Lucknow University, MEd.). Married Dr. Makhan Dubé of Princes Town. Emigrated to Canada in late 1960s, teacher and community leader in Nanaimo, British Columbia.
22. Vidiadhar (previously Vidyadhar) Surajprasad Naipaul, born 1932; son of Seepersad Naipaul, former journalist, *Trinidad Guardian*. Educated Queen's Royal College, Port of Spain, Island Scholar 1950; and University College, Oxford (BA English, 1953) (Carlton Comma (ed.), *op. cit.*, 1966, 178–9). Trinity Cross, 1989. Knighted in United Kingdom, 1990. Writer, greatly honored, and winner of the Nobel Prize for Literature, 2001 (Kris Rampersad, *Finding a Place: IndoTrinidadian Literature*, 2002; Patrick French, *The World Is What It Is: The Authorized Biography of V. S. Naipaul*, 2008).
23. H. V. Gopaul, born Princes Town, 1905. San Fernando merchant and real estate developer. Identified with social and charitable work ("Who's Who," Murli J. Kirpalani *et al.* (eds.), op. cit., 1945, 13).
24. Twenty out of 223 Hindus surveyed in San Fernando did not know their caste; the ratio was two out of 152 in Débé (Colin Clarke, *op. cit.*, 1986, 89).
25. Arya Samaj is a Hindu reform movement, and is anti-caste and anti-idolatory; but, like many Christian Indians, some Arya Samajis are caste snobs.
26. Not Hansar Ramsamooj. *Obeah* is black magic (Creole); an *ojha* man is a Hindu practitioner of black magic.
27. *Jajmani* is a system of economic exchange between occupational castes.

28. Eleven percent of San Fernando's Hindus and 9 percent of Débé's had participated in Siparu Mai (Colin Clarke, *op. cit.*, 1986, 111).
29. Pundit Jankieprasad Sharma, born in India in 1893. A Sanskrit scholar, he taught many pundits in Trinidad, and was versed in the Bhagwata Purana ("Who's Who," Murli J. Kirpalani *et al.* (eds.), *op. cit.*, 1945, 165; Kris Rampersad, *op. cit.*, 2002, 50).
30. A favorite Creole name for Indian wedding food of pumpkin and *channa* (chickpeas) is "lagoon fowl."
31. Om is the word which, when pronounced, has no end—implying eternity. It was introduced to Trinidad by Arya Samaj.
32. Ram Naumi is the celebration of Rama's birth.
33. Diwali, the festival of lights, is devoted to Lakshmi.
34. In the period before World War II, the Bhagwat was one of the few forms of congregational worship among Trinidad Hindus (Steven Vertovec, *Hindu Trinidad: Religion, Ethnicity and Socio-Economic Change*, 1992, 164–5). The Bhagavata Purana is related to the *Ramayana*, and treats Rama and Krishna as incarnations of Vishnu (Kris Rampersad, *op. cit.*, 2002, 32–3).
35. Mantovani was the conductor of a British string orchestra, popular at that time for its romantic and sentimental music.
36. Dr. Sarvepalli Radhakrishnan, born near Madras, 1888. Leading Indian of his generation in the fields of philosophy and religion; president of India, 1962–67.
37. Bhadase Maraj was out of national politics in 1964.
38. Andrew Moonir Khan, born 1905. Educated Couva, and Naparima Training College, San Fernando. Head Teacher of numerous Presbyterian primary schools, including San Fernando, 1956; Inspector of Schools 1956–62; Senior Inspector of Schools, Ministry of Education and Culture 1962– (Carlton Comma (ed.), *op. cit.*, 1966, 145).
39. Dr. Stella Abidh, daughter of Clarence C. Abidh of Charlieville, Chaguanas; attended Naparima Girls' High School and St. Joseph's Convent, San Fernando; Jarvis Collegiate Institute, Toronto (1921), and the University of Toronto. She was one of the first East Indian women to be educated to university level in medicine ("Who's Who," in Murli J. Kirpalani *et al.* (eds.), *op. cit.*, 131; Kris Rampersad, *op. cit.*, 117–18). Chaconia Medal, Gold, 1988.
40. Fears about communist infiltration were expressed in 1961, when the security forces searched the offices of trade unions and suspected individuals, such as C. L. R. James, the noted Trinidad Marxist and former teacher of Eric Williams (Selwyn Ryan, *Race and Nationalism in Trinidad and Tobago: A Study of Decolonization in a Multiracial Society*, 1972, 231).
41. V. S. Naipaul, *The Mystic Masseur*, 1957.
42. However, Steven Vertovec has disputed the supposed link between Kali Mai and Siparu Mai, and argued convincingly that Siparu Mai "is a goddess manifestation unto herself," *op. cit.*, 1992, 220.

43. *Maljeu* is French Creole for the evil eye. Lise Winer, *Dictionary of the English/Creole of Trinidad and Tobago* (Montreal, Kingston, London, and Ithaca: McGill-Queen's University Press, 2009)
44. *Jumbie* is Creole for spirit; ibid.
45. Dr. Eric Williams, the prime minister, was black-East Indian, or a *dougla*. *Douglas*, though racially mixed, are usually part of the Creole segment of Trinidad society.
46. Hans Guggenheim, an American graduate student in Anthropology, was, at the time, carrying out research on art and nationalism in Trinidad.
47. Andrew Carr, born 1902, recipient of Fulbright travel award to Northwestern University to study folklore, 1953. Leading ethnographer of Creole Trinidad; see, among others, "A rada community in Trinidad," *Caribbean Quarterly*, 3, 1, 1953, 35–54. Director, Little Carib Theatre; former director PNM Publishing Company; former President, Trinidad Art Society; member Carnival Development Committee (Carlton Comma (ed.), *op. cit.*, 1966, 67).
48. These Hindu forenames were never used, to our knowledge, in a home context.
49. The wife of Lionel Frank Seukeran, Member of the House of Representatives for Naparima.
50. Racial exogamy was rare for Muslims (2 percent) and Hindus (3 percent) in San Fernando; nil among Hindus in Débé. Only among San Fernando's Creoles and Christian East Indians did it reach 11 percent (Colin Clarke, *op. cit.*, 1986, 129).
51. Between 1952 and 1956, the Sanathan Dharma Maha Sabha built 31 schools around the island, and numerous temples were constructed or affiliated as well. The Maha Sabha published literature to be used in both sets of institutions (Steven Vertovec, *op. cit.*, 1992, 122).
52. Clarence C. Abidh, schoolmaster and sugarcane proprietor; father of Dr. Stella Abidh; contested Legislative Council seat for Caroni; lay preacher and member of Canadian Mission Board; first vice president, All Trinidad Sugar Estates and Factories Workers' Trade Union ("Who's Who," Murli J. Kirpalani *et al.* (eds.), *op. cit.*, 1945, 131).
53. C. L. R. James, born 1901; educated Queen's Royal College; taught Queen's Royal College, Pamphylian High, Government Training College for Teachers; Marxist freelance journalist, writer, and lecturer in England and the United States, 1932–58; editor PNM's the "Nation" 1958–61; secretary, West Indian Federal Labour Party (Carlton Comma (ed.), *op. cit.*, 1966, 133). Cofounder, with Adrian Cola Rienzi, Makhan Dubé, and Stephen Maharaj, of the Workers and Farmers Party; defeated in the general elections of 1966. He remained prominent in radical circles in the Caribbean and the United Kingdom, and was internationally recognized for his writings on cricket and Marxism. Trinity Cross, 1987. Died, 1989.

54. Duncan Village is located immediately south of San Fernando on the Débé Road.
55. Ralph Soodeen, born Princes Town, 1896, wholesale provision merchant and proprietor. Educated Naparima College; sportsman and racehorse owner. Lived in San Fernando ("Who's Who," in Murli J. Kirpalani *et al.* (eds.), *op. cit.*, 1945, 167).
56. Partition of Trinidad was proposed by the Indian National Association in 1962, during the debates leading up to the drafting of the independence constitution (Selwyn Ryan, *op. cit*, 1972, 330–1).
57. Creole heads of household who "married" East Indians (11 percent) usually chose Christians, and engaged no more frequently in "marriage" with other races than their Christian East Indian counterparts (Colin Clarke, *op. cit.*, 1986, 132.)
58. Kamaluddin Mohammed, born 1926, former radio announcer. Elected PNM Member of the Legislative Council for St. Joseph 1956–61, Member of the House of Representatives for Barataria, 1961–; Minister of Agriculture, Lands and Fisheries, 1956–61, Minister of Public Utilities, 1961– (Carlton Comma (ed.), *op. cit.*, 1966, 175). After 1966 he went on to hold ministerial posts in all the Williams administrations, but was overlooked when Williams died in 1981, and George Chambers was asked to form a government (Hamid Ghany, *Kamal: A Lifetime in Politics, Religion and Culture*, 1996).
59. Dr. Winston Mahabir, born 1921, son of Jules Mahabir. Educated McGill University, Canada; physician and surgeon. Elected Member of the Legislative Council for San Fernando West, 1956–61, Minister of Health, 1956–61. Emigrated to Vancouver, Canada (Carlton Comma (ed.), *op. cit.*, 1966, 163). See Winston Mahabir, *In and Out of Politics*, 1978.
60. Traditionally, Hindu marriages were arranged by an *aguwah*, though we did not come across anyone in our research playing such a role.
61. Brahmachari Hari Ram appears in Morton Klass, *op. cit.*, 1961, 150.
62. James Sammy, born San Fernando, 1890, educated Naparima College, San Fernando and Dalhousie University. Teacher, Naparima College; secretary Naparima College Board and Susamachar Board of Control ("Who's Who," in Murli J. Kirpalani *et al.* (eds.), *op. cit.*, 1945, 163).
63. Dr. Makhan Dubé, born Princes Town; M.D. and general medical practitioner, Princes Town. Married Vilma Meghu. Cofounder, with Adrian Cola Rienzi, C. L. R. James and Stephen Maharaj, of the Workers and Farmers Party; defeated in the general elections of 1966. Emigrated to Canada in late 1960s. Medical practitioner and community leader in Nanaimo, British Columbia.
64. Hosay is the festival celebrated by Shiites to commemorate the martyrdom of Hosain, Mohammed's grandson, in AD 680. Virtually all Trinidad Muslims are Sunni, and the celebration of Hosay is largely confined to St. James, Port of Spain, where Creoles also participate.

65. Syed Mohammed Hosein, born 1888; licensed Hindi interpreter and Muslim divorce officer; founder of Anjuman Sumat al Jamaat ("Who's Who," in Murli J. Kirpalani *et al.* (eds.), *op. cit.*, 1945, 141).
66. Both the ASJA and TML mosques in San Fernando are breakaways from the Takveeyatul Islamic Association (TIA) founded by Abdul Aziz of Iere Village, Princes Town in 1926. The more conservative ASJA was incorporated in 1935, the TML in 1947. The TML approves of equal rights for women, includes them in the congregation and other meetings (the ASJA segregates its congregation and screens off women participants), counts eight rather than twenty *rakaats* in a prayer cycle, and believes, as ASJA does not, that Mohammed is the one and only prophet (Colin Clarke, *op. cit.*, 1986, 101).
67. *Tadjahs* are wood, paper, and tinsel representations of the tomb of the martyr Hosain, constructed for Hosay.
68. Just over half the Muslims in San Fernando and Débé fasted for Ramadan, and almost all celebrated Eid ul Fitr and Bakra Eid (Colin Clarke, *op. cit.*, 1986, 108).
69. Jaleel Sheik Mohammed, born 1895; aerated water manufacturer, patron of Oriental Cricket Club, member of Anjuman Sumat al Jamaat ("Who's Who," in Murli J. Kirpalani *et al.* (eds.), *op. cit.*, 1945, 151).
70. Ranjit Kumar, an Indian (see also Part 3, footnote 64) is also credited with the import of the first Hindi film—*Bala Joban*, which ran at both the Globe Cinema in Port of Spain and the Gaiety in San Fernando (Lynne Macedo, *Fiction and Film: The Influence of Cinema on Writers from Jamaica and Trinidad*, 2003, 33).
71. In the 1950s "Jangli Tola" was occupied by crab catchers and fishermen (Morton Klass, *op. cit.*, 1961).
72. Joint households are rare among Hindus in San Fernando, but more common in rural areas. In the Débé survey, for example, one-third of households had three or more generations living together, compared to just over one-fifth in San Fernando. However, in both town and country, Hindu brothers often live in close proximity, and it is common for their nuclear households to collaborate (Colin Clarke, *op. cit.*, 1986, 117).
73. Saied Mohammed, born 1916; civil servant and accountant in private sector. Member of the House of Representatives 1961–; Minister of Local Government and Community Development 1961–64; Minister of Works 1964–; former Vice President Anjuman Sumat al Jamaat (Carlton Comma (ed.), *op. cit.*, 1966, 175).
74. Tajmool Hosein, born 1921, educated London University, BA. and LL.B. (Hons.); called to the Bar, Lincoln's Inn, 1946. Elected DLP Member of the House of Representatives 1961; 1963 left the DLP (at same time as Peter Farquhar) to form short-lived Liberal Party with white support (Carlton Comma (ed.), *op. cit.*, 1966, 125). Queen's

Counsel, Trinidad and Tobago, 1964; Queen's Counsel Caribbean Associated States, 1982; Trinity Cross, 1982; Hon. LL D University of the West Indies, 2005.

75. Presumably, the San Fernando Hindu Sabha was associated with the Prince Albert Street temple (Steven Vertovec, *op. cit.*, 1992, 117).
76. Trinidad French Creole for teasing, ridicule, or an insult. See Lisa Winer, *op. cit.*, 2009.
77. The Canadian Mission of Rev. John Morton was founded at Iere (now Princes Town) in 1868, and moved to San Fernando three years later (Sarah E. Morton, *John Morton of Trinidad*, 1916).
78. Indian indenture (started in Trinidad in 1845) was abolished by the Indian government in 1917, but as the term was five years (for men), indentured labor did not disappear from the colony until 1922.
79. *Who's Who in Trinidad and Tobago, 1966* showed that 72 percent of San Fernando's elite (63 persons) were Creole (14 percent women); 28 percent were East Indian (2 percent women). Among the East Indian elite, Christians outstripped Hindus by two to one, and no Muslims were listed. However, it is likely that editorial bias toward formal qualifications undervalued the Hindu and Muslim Indian contributions to public life (Carlton Comma (ed.), *op. cit.*, 1966).
80. The names Coombs and Gibbon are mentioned specifically in Mariane Ramesar's chapter "The Repatriates," in David Dabydeen and Brinsley Samaroo (eds.), *Across the Dark Waters: Ethnicity and Indian Identity in the Caribbean*, 1996, 175–200.

Part 3

Conversations

Saturday, July 4

Gillian: *Mrs. Rampersad on sanskar*

We had a long discussion with Mrs. Rampersad at her home in San Fernando about *sanskar*, or Hindu life-cycle celebrations.[1] Just before the birth, prayers are said for a healthy baby. After the birth, it is important to give thanks. If the baby is strong, its head should be shaved on the seventh day; if the baby is weak, the shaving can be postponed or not done at all. A second ritual shaving takes place at one year.[2] There is no need, in Trinidad, for the Nau to be of the barber caste, as long as he is a barber.

The father may not see the baby immediately after its birth. First, the priest must make an astrological reading and decide which day is suitable. When it is appropriate for the father to see the baby, its face, reflected in oil in the base of a *tarriah*, is shown to him. *Kajar*, or soot from the burning of *ghee* or clarified butter, is placed on the baby's eyelids during its first six days—to clean the eyes. Jet beads are often put on a child's wrist to protect the child from the bad looks (*maljeu*) of people, particularly of barren women. (Mrs. Rampersad does not believe in this, and never put beads on her children.)

On day 6 (*chati*) or 12 (*barahi*) after the birth (the priest decides which one), the nama sanskar takes place. The barber comes and trims the nails before the mother and child are bathed in the morning. A prayer is offered to the sun, and the bathwater is thrown away toward the sun in the east. Both mother and child are dressed in new clothes. Friends and relatives bring gifts, and there are prayers with cooking and singing in the evening—sometimes drinking. The father will be present, but will not see the baby unless the priest has given his approval. The treatment is identical for boys and girls. Mrs. Rampersad's friends still maintain these rituals, but she thinks the younger generation will find them old-fashioned.

Boys should be baptized at seven years of age, but this is now done mostly by Brahmins. Gurudiksha involves the equivalent of a christening and the choice of a preceptor. One can do gurumukh on the same day as a family *katha*, and the *guru* (godfather) can be any pundit. The rituals involve bathing, prayers, the construction of a *vedi* with a fig tree and nine elements represented by nine flags. Girls do gurumukh, but don't take the *janeo* (sacred thread), unlike boys, for whom the ceremonies are often on the same day.

The *janeo* should be taken by young Maharaj men before marriage, since it confirms Brahminical status on them. Saffron is worn for several days, and ritual bathing, nail cutting, and shaving take place, accompanied by prayers. The family call their pundit and friends for the occasion. The boy dresses like a hermit, with a *dhoti*, and wears a cloth around his shoulders like a beggar. The women offer him small gifts, which he collects and presents to the Nau and the pundit. He must be considered on that day as a saint, not as an ordinary human being. His head is shaved in five sections, the first time after a reading, and the second after another reading. Sometimes the head is shaved in one go, but a *churkee* (topknot) is always left.

The *janeo* ceremony takes place in a *maro*, with a *vedi*, very much like the Hindu wedding. But the boy makes no special promises, other than reaffirmation of respect for his parents. Some boys take the *janeo* before they leave Trinidad to study; many parents are keen that boys should take the *janeo* in case they leave the island and get married.

Girls may take the *oronhi* at any time, but it is intended to mark the onset of puberty. Nowadays many don't bother. "We livin' in a mixed world here. We not in India."

Sunday, July 12

Colin: *Walter Annamunthodo on religion and politics*

I had arranged an interview with the trade unionist and publisher Walter Annamunthodo,[3] a Madrasi. His father was Hindu, his mother Christian, and he and his brother, Harry (the surgeon at the UWI hospital in Kingston, Jamaica), were born and raised in British Guiana. Walter was engaged in trade union activity in Aruba in the 1940s, but had been repatriated to British Guiana, where he was drawn to the Arya Samaj (which permits conversion).

He commented on the uplifting work of the Presbyterian Church in San Fernando, and its role in producing professional and white-collar men—Wilson, Lutchmansingh, Samaroo, Sumadh, the Sinanans, and the Mahabirs.

Arya Samaj is one of the shareholders of the Vedic Press of San Juan, in which Walter is involved. He has a long history of association with the Oilfield Workers Trade Union, the West Indian Independence Party,[4] and with C. L. R. James. In 1961 he took over management of the Vedic Press, and published James's book *Party Politics in the West Indies.*

When the PNM was formed and swept to power in 1956, one of the successful PNM candidates was Dr. Winston Mahabir, who took San Fernando West. Walter claimed that Mahabir drew on a great deal of East Indian support, not because he was Indian, but because the Indians understood the issues of the day. Although Mahabir was not an active churchman, the Presbyterians gave some support to the PNM. His opponent, the old-time politician, Roy Joseph,[5] though standing as an Independent, was backed by Bhadase Maraj and the orthodox Hindus, with the Arya Samaj siding with the PNM. Joseph (of Indian/Syrian origin) was married to a Muslim, but in San Fernando, the Muslims were politically split. Although he did many favors for the Muslim community, he lost.

D. L. P. Convention, Naparima Bowl

Later in the morning, I attended the DLP Convention[6] at the Naparima Bowl in San Fernando, accompanied by Bram. The rules of the party under democratic socialism were read out by Mr. Khan, a master at Presentation College. One man commented, "If the leader is satisfied with the rules, then we shall be satisfied, too," but he was shouted down. It was stated that the membership of the party was 20,000–30,000, and that it was to work toward creating a multiracial society. I noticed Alloy Lequay[7] and Clive Phil among the delegates. Also present was the journalist H. P. Singh,[8] whose wife is H. V. Gopaul's sister. On the stage, I counted 54, of whom 49 were East Indian, 3 white and 2 Chinese; among the remainder, which I estimated at about 450, there were 400 East Indians, 42 blacks (all middle-aged) and 5 whites.

Among the delegates I identified: Jang Bahadoorsingh, Simboonath Capildeo (not on speaking terms with Rudranath), Ram Surasingh, and Chanker Maharaj[9] (the former wrestler). Bram tells me that Mr. Gopie was once a "big man" in the Maha Sabha, but there seems to have been rivalry with Simboonath. Gopie says he is not prepared to beg for favors; he has "clear" skin, not black (like Simboonath). Dr. Avatar is the district medical officer for Moruga. It is said that Simboonath Capildeo wants to take control of the Maha Sabha, using Jang Bahadoorsingh as a front man, and then grab the DLP.

It seems that the old DLP Executive had threatened to oust Dr. Rudranath Capildeo, unless he returned to Trinidad to live. This Executive, which included many businessmen, including De Lima, was eventually sacked by Capildeo. The outgoing Executive was apparently largely from Port of Spain.[10] Now the proposal is to have members elected from each constituency, of which there are 30 in total. Capildeo himself noted that, as the Executive would be elected in a public assembly, there would be no excuse for "knives in the back."

Someone said of Capildeo, the leader of the opposition, that "he enjoys the respect and confidence of the country." But a black delegate commented that there was too much sentiment about Dr. Capildeo, and not enough thought being given to the definition of the role of the party leader. It struck me that the black delegates were hardened politicians opposed to Williams and the PNM, probably because of some personal disagreement or political slight. One of the major outcomes is that Dr. Capildeo is clearly going to continue with his post at London University. In his closing address, Capildeo signaled the "end of the era of the educated idiot" (meaning Dr. Eric Williams).

Monday, July 13

Colin: *Dan van Eendenburg on caste and marriage*

Dan van Eendenburg underlined many of the conclusions we have been coming to. Caste can be circumvented by such things as color. For example, a pundit told Dan that he was marrying his daughter to a Chattri boy—who was very fair-skinned. Girls are still forced into marriage. A brother was determined to marry off his sister of 20 when his father died. She threatened to run away, but later agreed to marry the older man chosen for her. But she cried and refused to eat for two weeks before the wedding. Now her husband is already beating her, and they have been married for less than two years.

Hansar Ramsamooj on Brahmins in San Fernando

In a later conversation, Hansar said that the Brahmins in San Fernando use him because they cannot get a pundit to take the Sunday-morning worship. They refuse to give him financial support for SSDM, though they do help the Divine Life Society of Brahmachari Hari Ram. Bissoon Pundit's verbal attack on him (for acting as a priest without being a Brahmin) was followed by a very successful *hawan* Hansar held in the San Fernando home of a young Hindu doctor who had recently returned from Canada. Hansar commented that it is ironic that he (as a non-Brahmin) spends every Sunday morning teaching a Brahmin-dominated group of children.

The temple congregation is predominantly Brahmin, according to Hansar, largely because Brahmins are overrepresented in San Fernando. Furthermore, the Brahmins have the temple in their pocket, and others feel they are not welcome. Mrs. Maharaj, the Arya Samaji widow of Suruj Balli, does not attend because she feels an outsider. But Hansar is to some extent breaking into the charmed Brahmin circle. At the end of June, Mr. Deabi Persad Maharaj held the first *katha* he had ever given, officiated by Karsiprasad from Tunapuna. At the dinner were Karsiprasad, Jankieprasad Sharma, Bram, Gopaul-Gosine, Binie Maharaj, Gopaul, Gayah, and Hansar. Although pundits are always polite to Hansar to his face, they later indulge in backstabbing. Jankieprasad Sharma is said to be hostile to him.

The Brahmins in San Fernando are a religious clique. Their power of exclusion probably affects Hansar more than it affects anyone else. It takes a great deal of pushing from Bram to get Mistress Deabi to give Hansar some shirts as a token of their so-called appreciation. On the other hand, Ramnarine, who is wealthy but a Chamar, likes to hear Hansar precisely because he is not a Brahmin. Hansar is the *chela* (initiate) of a *swami*, not of a Trinidad pundit. Hansar was christened privately at home, and given his *mantras* by his *guru*. Hansar has not taken the *janeo*, though he says that any Hindu can do so. He thinks that he cannot live up to the standards required. He knows a *pundit* in Trinidad who would give it to him.

Trinidad Brahmins are annoyed that certain important *mantras*, regarded as the private property of Brahmins, are included in the sandhya puja used by SSDM. Brahmins do the *hawan*, but are never taught *mantras*, unlike the Arya Samajis. Hansar was once attracted to Arya Samaj, but the local Arya Samaj pundits, Vishnudath at Diamond and the late Suruj Balli in San Fernando, were both Brahmin. Hansar decided that there was nothing in Arya Samaj for him (as a leader, I assume).

Bisram Bissoon on caste and marriage

Bisram Bissoon, who assists Hansar in the *hawan*, would like to be a pundit. He has been taught by Hansar, whom he calls Fa, and (so he says) by Mr. Bhattacharya. However, Bisram is a Kurmi,[11] not a Brahmin.

Bisram does not think that caste is important, though Brahmins in Corinth have objected to Chamar and Muslim men preparing marriage food (they were soon shouted down). But Bram cooked for Ramnath Maharaj, and Hansar was not even invited. Bisram cooks and makes *roti* at weddings, having been taught by Mr. Ramjohn, the estate foreman at Usine. Bisram is pleased that, although his sister's policeman husband is a Chattri, they have never made him feel that he is of lower caste, nor have they openly enquired of the family's caste. But Bisram does not doubt they know it.

Bisram's father met his future son-in-law, conversed with him, and found out that he was a Singh and wanted to get married. Bisram's sister was introduced to him, and a marriage between the two was agreed upon. The families visited one another. The Bissoon family gave the groom a cow and about BWI$400 (£80). Officiating at the wedding ceremony was split between Hansar and a pundit, because Hansar is neither a Brahmin, nor licensed to perform marriages. The marriage was registered, and a *puja* was held in Tunapuna to welcome the bride and groom. All the Singh brothers live in the same street, on land owned by their father.

Bisram harbors hopes of getting married soon. Friends are on the lookout for a girl for him; so is his brother-in-law in Tunapuna. Bisram does not mind about caste, but he would prefer a light-skinned wife. She must be a good girl, not with a secondary education, but able to cook, sew, and look after a home. The domestic qualifications are important, because Bisram's mother works in the cane fields, and it is hoped that her eldest daughter, who is the family's housewife, will soon be married. Bisram's future bride will have to assume this daughter's role within the family.

Bisram has seen two girls already, one of whom lives in Corinth and is the only girl in the village he would like to marry. But she is going in for secondary education, and Bisram does not think she is worth waiting for. Besides, Bisram is loath to take a bride from the same village,[12] because it means having too many relatives in the vicinity. In the case of the second girl, he did not like the look of her, her home, or her parents, so he did not even talk privately to her. He sent word to say that he was not ready to marry yet.

Mr. Gopie on caste and politics

Mr. Gopie told me that he does not attend the temple on Sunday mornings because he has more pressing practical work to do. He is associated with welfare in the sugar industry, with fishing, the Gandhi Seva Sangh (Service League), and music groups. He is outspoken against caste, which he claims is increasing, not decreasing in importance in Trinidad. In India there has been official support for the breaking down of caste, but not here; Brahmins have actually raised barriers. It is because of caste that he is in the PNM camp.

Lionel Seukeran, the DLP representative for Débé, has done precious little to help his Naparima constituency.

I was invited to a PNM meeting (Group 16) in Mr. Turton's house in San Fernando. About 17 people attended, including Mr. Davis, the constituency chairman, Mr. Gopie, the chairman of Group 16, Western Constituency, Errol Mahibir,[13] Mayor of San Fernando, and his father-in-law, Alderman Lans. Of those who were present, nine were black, five East Indian, and three neither black nor East Indian. Interracial solidarity was stressed during the meeting. Mr. Gopie enumerated the work he had been doing among the Indians and reminded everyone that he is president of the Indian Musicians. This will, he says, enable him to promote the policy of the PNM in this organization and to infiltrate the East Indians more generally.

Thursday, July 23

Colin: *Rev. James Seunarine on education and Christianity*

During the evening I had an interview with Rev. James Seunarine,[14] headmaster of Naparima College. He has been head of this Presbyterian secondary school for three-and-a-half years, and before that was in charge of the Presbyterian Theological College (which is largely involved in lay-leadership training). Rev. Seunarine's grandfather had been the caretaker of Paradise Pasture, where the college is located.

Rev. Seunarine's father had trained as a pundit, and, after conversion, as a Presbyterian catechist. He had planned to go to India but gave up the idea, and went to work in the oilfields after his wife died. James Seunarine won a bursary to Naparima College, and helped his widowed maternal grandmother clean the school windows to make ends meet. Like other Indian boys of his age, he had been educated at primary level in CMI schools, where the "I" was for Indian (Canadian Mission Indian). Presbyterian Indians were a distinct group, but by no means cut off from Hindus. The missionary impact of the Presbyterians is much less than it was 40 years ago. The 1930s depression led to a reduction in personnel being recruited to the church, and there are few ministers in the 35–60 age bracket.

Why did Hindus convert? There were educational and occupational benefits, especially in teaching. Despite current problems, there are quite a few Presbyterian ministers who converted from Hinduism and Islam, though, in general, Muslims have been more hostile to Christianity than Hindus. According to Rev. Seunarine, there is "no definitive situation" with regard to caste and conversion; converts were neither all low nor all high caste, though some Christian Indians claim that "caste always shows out," especially in breeding. Hindus who became Christian were often referred to by other Indians as "creolized," while Creoles described Susamachar as the "coolie church."

The American-based Church of the Open Bible has recently made quite an appeal to Indians, but it has preferred the "sheep-stealing" of Presbyterians to conversion of non-Christians. The Church of the Open Bible appears to provide a sense of fellowship missing in the Presbyterian Church; it is more spontaneous and evangelistic. Presbyterianism is heavily institutionalized and very deeply committed to education, yet it is losing teachers to other professions, such as the Probationary Service.

The feeling of being an "Indian" is very strong in Trinidad. One is "Indian" first and "Christian" second. Intermarriage between Indians of different faiths is, however, a difficulty. "Even those who accept, do not do so happily." Rev. Seunarine was, as a young man, engaged to a white Canadian girl, but she married a Canadian missionary. Rev. Seunarine is in favor of mixing, though as a Trinidad Indian, he would prefer his child to marry its own kind, but that is merely a preference.

Hindu youth, in Rev. Seunarine's eyes, are "lost people," and probably less moral than Christians nowadays. Even in the country, close social connections between people are falling apart. In town, people lead socially isolated lives. Indians and blacks are often close at school, but do they carry this into later life? The more sophisticated they are, the more they move apart. Rev. Seunarine's closest friend at school was Pelham Sloane-Seale,[15] a Creole Chinese.

Saturday, July 25

Colin: *Mr. Bhattacharya, Tagore College, and Brahminism*

Mr. Bhattacharya declares himself disillusioned with the East Indians, and especially the Brahmin leadership. When he first came to Trinidad, he was warned by the big boys—especially Gopaul—to keep away from Mr. Ramsamooj. The money making of the East Indians makes Mr. Bhattacharya sick. Moreover, caste, especially Brahminism, is very strong. Businessmen will not go to the Todd Street *mandir* because they say it is in the pocket of Bisram Gopie and the PNM. Mr. Bhattacharya claims that is the reason they say they will not go: but even if Mr. Gopie pulled out, they still would not go. Bombay women conduct their own *puja* in the Todd Street *mandir* during the Sunday Sandhya Puja, thus disrupting Hansar Ramsamooj. Mr. Bhattacharya claims that most Hindus in San Fernando are Brahmins.[16] In his opinion adultery for Hindus is not a moral issue involving sin, but a social problem.

Mr. Bhattacharya had hoped to go on leave to India this year, taking some of the leading Trinidad Indians to show them the country, and indicating what he could achieve for them. They refused, saying that they knew all there was to know about India. Mr. Bhattacharya is in a difficult position, since he has been brought to Trinidad by Jang Bahadoorsingh, Gopaul, Seunarine, and Mr. Gopie.

Tagore College has been created out of the overseer's house on Gopaul's 600-acre estate at Craignish outside Princes Town, and Gopaul has footed the repair bill of BWI$40,000 (£8,000). Jang Bahadoorsingh pulled out early on, leaving Gopaul to finance the project—and he has been trying to turn a profit from it! George Sammy thinks that Gopaul wanted the school so that he could dispossess the farmer-tenants who were living on his land. Giving the land to the school was an excuse; the real objective was the creation of another Marabella Lands subdivision. Despite Mr. Bhattacharya's gentleman's agreement with Gopaul, the latter is not interested in Indian education, but just in maneuvering.

In Mr. Bhattacharya's view, Brahminism is not being broken down at all. To be a Brahmin is a special cachet for all those who have made it socially. Mr. Bhattacharya hinted that many Brahmins are not genuine, and I think Gopaul could be one of them, from what was insinuated. To be a Brahmin is to have status. A mother might try to get her son married to a Brahmin by claiming that, although she is not a Brahmin, she committed adultery with a Brahmin, and that the boy is consequently a Brahmin. Her shame in adultery would be cancelled out by his pedigree.

Mr. Bhattacharya's name means *guru* of the Brahmins. Ramnath Maharaj's daughter, of whom Mr. Bhattacharya seems very fond, asked his advice, as a *guru*, before she agreed to marry. The marriage ceremony in Trinidad has been standardized by Brahmin priests who knew only the orthodox form of ceremony. Caste variations in India depended largely on caste priests, according to Mr. Bhattacharya.

Peter Dubé has told us that he thinks Gopaul wants to give Mr. Bhattacharya the push. He offered Mr. Bhattacharya three months' "pay" and passage to India for him and his family. But

Mr. Bhattacharya gave up everything to go to Craignish, and he will feel that he has failed if he leaves now. A Mr. Persad from San Fernando has offered to put up new buildings and finance a new Tagore College.[17]

Seunarine puja at Lothian Estate

While Gillian and Ena went to Port of Spain during the evening, I attended a Seunarine *puja* at Lothian Estate, south of Princes Town. The house was very simple on the outside, but the setting inside of the specially constructed "tent" was particularly rich, notably the canopy under which the *mahants* (low-caste priests) sat. The *chauk* (altar) was covered with a pure white cloth, on which stood two vases with paper flowers, and three *dias*, resting on rice-filled cups. The *mandir* was full of beautiful red and yellow colors—partially opened hibiscus and marigolds—while tinsel and streamers made of flowers decorated the canopy.

Men and women were segregated at either end of the tent, except for the householder and his wife, who sat close to the officiating *mahant*, just outside the central canopied area with the *chauk*. Prayers were said with hands on lap. Then there was worship of Swami Narayani's Guru Anyas (a compilation of 16 holy books), and black ointment put on top of an open page, followed by the frenzied singing of *bhajans*. The officiating *mahant* and his assistants placed a white thread around the open Guru Anyas. A fire was lit, and the host and his wife joined in the ritual. Small brass *tarriahs* with *pan* (betel) leaf inside were filled with coconut and sweetmeats and placed on the *chauk*. *Pan*, I was told, is chewed by people from Madras. Mr. Rampaul, president of the Trinidad Seunarine Sabha, pointed out to me that no rum or other alcohol was used.

Prayers were chanted, and *aarti* was again performed by the husband and wife, followed by other participants. Many of the women wore multiple silver bracelets. The *mahants* performed *aarti*, and I was asked to join in. During *aarti*, the same song was sung as among Sanathanists. *Aarti* was performed by four hands, involving a couple, and then again by the *mahants*. *Aarti* was performed over a picture of Seunarine (Swami Narayani) by the hosts, followed by the blowing of the conch, the ringing of the bell, and the chanting of Seunarine Swami Ki Jai. However, rum was being drunk by the musicians, who had a harmonium, *tabla*, and *manjira* (cymbals), and by some of the women at the back.

The husband sat crossed-legged before his wife and received a *tika* on his forehead from her. Prayers were said to Guru Seunarine, and those near enough touched the *chauk*. The girls began singing *bhajans*. Two additional *dias* were lit using oil, not *ghee*, and the husband and wife put incense on the fire as an offering. *Persad* was placed in a *tarriah*, with banana leaves and hibiscus. *Aarti* was done with fire and incense over the *persad* and the *chauk* by one of the *mahants*, Janglie Ram, who was described to me as the Seunarine pope.

The *persad* was placed touching the *chauk*, and *tulsi* leaf was thrown on the *persad*, which was added to five small *tarriahs* full of sweetmeats. Water was poured on the *bedi* from a broken mango leaf, followed by the ringing of the bell. A white cloth was placed on the Guru Anyas, and it was sprayed with perfume—"the spice of God." Prayers were said and everyone touched the *chauk*. Water was dripped from the mango leaf into the *tarriah*, in worship of the Guru Anyas, and the water was drunk by the wife.

Mr. Rampaul said prayers in front of the *chauk* and coins were offered, and this act of worship was repeated by other men. Closing prayers were said to all the *deotas* (gods). Incense was offered to the fire, prayers were said to the *deotas*, and *persad* fed to them. *Aarti* was performed for the last time by the husband and wife, accompanied by the bell, conch, and gong. Prayers were chanted in unison. The ceremony came to an end, but the *mahants* literally could not move until the canopy over the *chauk* was taken down.

After the ceremony, I drove back through the night on my own to San Fernando and then on to Port of Spain, where friends of Ena had—allegedly—planned a Sunday trip "down the islands" toward Venezuela. Unfortunately, no one had thought to book this outing in advance, and all the boats were taken. Ruefully, we returned to San Fernando on Sunday afternoon.

Monday, July 27

Colin: *Dharam Pundit's Durga puja*

I went in the morning to Diamond Village to visit Dharam Pundit.[18] Here I witnessed the priest and his wife, known as "Christmas Tree," perform a Durga puja to protect their son's new house, on the neighboring lot, from sickness (Plate 10). On a previous visit I had participated in a ramayan satsang held at the pundit's house, during which part of Tulsidas's Ramayana had been sung by villagers. Dharam introduced me to Ramcharittan Pundit, from Lower Débé, who had been present for the Durga puja, but had been fed separately from me. Ramcharittan assumed that I knew nothing about Hinduism. He

Plate 10 *Durga puja* (ceremony devoted to goddess Durga) being performed by Dharam Pundit and his wife. Note the brass *lotah* (round vessel) and *tarriah* (circular tray), and the rum bottle containing *ghee* (clarified butter)

expressed himself as anti-PNM and antiblack. He is a great supporter of the UK Immigrants Act, not realizing that it is aimed at Caribbean Indians and blacks alike!

Hemant Kumar show

In the evening Gillian and I went to the Naparima Bowl to the Hemant Kumar show from India. With an audience of between 700 and 800, there were few saris but a mass of *oronhis.* The drummer was wonderful, though we rapidly tired of Kumar's Indian film "pops." The program was expensive, monotonous, and badly organized. But the sentimental Indians lapped it up and even packed out an extra night.

Tuesday, July 28

Colin: *The Church of the Open Bible*

Rev. Whitlow and Rev. Wood, both white Americans, talked to me about the founding of the Church of the Open Bible in Trinidad. The parent church is the Open Bible Standard Church Inc. of Desmoines,

Iowa, United States. In the early 1950s, the missionaries Karre and Jean Wilhelmsen were not allowed to return to India after spending their leave in the United States, so they looked for a Hindi-speaking community in need of the gospel, and chose Trinidad. He was Norwegian and she, American, and they started holding meetings in their home in San Fernando. The policy of the church is to avoid the principal city of any country.

The sick were brought to them and healed. Healing is the basis of their work. There are negative and positive sides to healing—black and white arts. "The world is hungry for a God who does what he says." "Faith is a dynamic, fear the reverse." Faith must be channeled: "Tongues we have never heard we understand." The Wilhelmsens stressed that "if a person takes Christ, everything else must go out. It keeps the work pure. We have nothing new; encounter the person of Christ."

The Wilhelmsens came for the Hindus, but soon other races were attracted, blacks in particular. However, the choruses are still sung in Hindi—"blacks love to learn and sing." Repeating the experience of the Mortons in the nineteenth century, the Church of the Open Bible has found that Muslims have remained aloof: "the only way we reach a Muslim is through a miracle." Some rural Hindus, however, have been hostile, and the Church of the Open Bible was for a time driven out of Débé, where it currently has two areas for services. "The blood of religious experience runs thicker than national and racial identity."

It is claimed that the Church of the Open Bible has about 500 members, with about 600 people attending the four Sunday services. There are also prayer meetings or services on four weekdays, including the Mountain Movers faith-healing service every Thursday morning. In five years' time, they hope the church will be run by Trinidadians.

Three-quarters of the work is with Hindus and Muslims, though others are well represented. It takes the form of self-generating missionary activity based on the congregation that attends. The pattern is one of attendance at church, the creation of a spiritual crisis, then relief through personal commitment to Christ. The deacons were originally rigged to be fifty-fifty black and East Indian; now free elections are producing the same ratio. Gillian and I had previously attended a Mountain Movers service, where the same pattern of sin/guilt/saved was evident. On that occasion 15–20 people stepped up to the altar at the bidding, "Many of you are here for the healing."

Rev. Ramjit, Susamachar Church

Later the same day I had an interview with Rev. Ramjit, the Presbyterian pastor of Susamachar Church. He is in his sixth year at Susamachar, and claims that more than 40 adult Hindus and Muslims converted to the Presbyterian Church during the previous year, despite the missionary appeal of the Church of the Open Bible. He himself was born and baptized an Anglican, but he speaks and reads Hindi, and implies that a Presbyterian pastor must have a good knowledge of Hinduism. He started as a catechist more than 30 years ago and eventually became a pastor.

He complains that young ministers now know only English, and have no background in comparative religion. In his view, Christ fulfills the deficiencies in Hinduism and Islam. Throughout Rev. Ramjit's incumbency of Susamachar, services have been only in English, but in the country, services are often partly in Hindi—"offering Christianity in their own vessel," as he put it. There has been some loss of congregation to the Church of the Open Bible; "some people like that brand of Christianity." The congregation at Susamachar was rather neglected before Rev. Ramjit took over. He finds that Muslims are more difficult to convert than Hindus, because they have no sense of polytheism.

The Presbyterian Church, in his assessment, is 95 percent East Indian. Although it tries to appeal to others, it has only a few blacks, which is "a weakness." Some East Indians still say of the congregation, "Why do blacks want to come here?" When Rev. Ramjit was in Roussillac, south of San Fernando, in 1933, the East Indians were opposed to blacks taking communion with them. But it was a tactical mistake of the missionary leadership to equate the Presbyterian Church with the "Indian Church."

Christian Indians are in a way acculturated, though they maintain Indian family values, sense of fraternity, and respect for the elderly, Indian food, and some other "indefinable characteristics." Presbyterians have a tendency to be insular, though in San Fernando, because of the Presbyterian Schools, Hindus and Muslims will mix with them. The concentration of Presbyterian primary and secondary schools in San Fernando has given Indians access to higher positions than they enjoy elsewhere.

Caste lives on in the hearts of some members of the higher castes who became Christian. At a Presbyterian Synod meeting, a Christian Brahmin commented, "You can always tell a low caste man by his deeds." Rev. Ramjit added, "There might be an element of truth in that." Christian Brahmins and Chattri still try to get high-caste spouses for their children,[19] and displeasure is often expressed if high-caste boys marry low-caste or Muslim girls. Muslims are associated with the PNM. Some Presbyterian parents still arrange marriages, in the expectation that parents, "being more mature," should be able to see the pitfalls for the young married couple—and this is sometimes a good thing.

In general, it is easier for Christian missionaries to convert members of the low castes, though in Trinidad, Presbyterian Indians are representative of the caste hierarchy, in contrast to the situation in India.[20] Some converts drop the names Maharaj and Singh to assimilate, though talk about caste might come up incidentally.

Wednesday, July 29

Colin: *Faizool Rahaman, ASJA General Board*

Faizool Rahaman is a teacher, and a major figure in the Muslim community in San Fernando and nationally. He is the acting chairman of the ASJA General Board, while the chairman, Ibrahim Mohammed, who is always in Port of Spain, happens to be away on *Haj* (pilgrimage to Mecca). The ASJA General Board looks after the ASJA Boys' and Girls' Schools, in the case of the latter via a subcommittee that deals with only certain aspects. The schools were opened in San Fernando four years ago (for the boys) and two years ago (for the girls), and they have enrollments of 500 and 250, respectively. In addition, ASJA Infants School takes more than 500 children. The Prince Albert Street Mosque (TML) is overcrowded, so the Friday lunchtime service for the infants is held in their school.

Prayer cycle

A few, including Mr. Rahaman, keep up the five daily prayers, either at home or in the mosque. The presence of the imam is not necessary in the mosque; if only three participants are present, one will take the lead. The five times for prayer are: shortly after midnight, midday, mid-afternoon, late afternoon, and last thing at night. If you miss a point in the sequence, you can read extra prayers to make up, but the maximum number of *rakaats* (genuflections) is 17.

A *rakaat* is measured in the following way: Stand to the creator; hands to ears; bow; hands to knees and up again; forehead, palms, and nose to the ground twice; finish with hands on stomach, and roll head to right and left. If you are traveling you can do your prayer positions with your trunk and eyes instead. Throughout, it is compulsory to face toward Mecca, and the mosques are carefully aligned in this direction.

Ramadan

The highlight of Muslim life in Trinidad is Ramadan, a 29- or 30-day fast depending on the lunar month. The mosque is overflowing every night, and the fast provides discipline and spiritual purification. Ramadan "helps to equalize the community"; sexual intercourse is forbidden during Ramadan; and the fast extends from dawn to sunset. During Ramadan Mr. Rahaman gets up at 4.30 a.m.—before dawn—and eats. This is followed by morning prayer and the reading of the Koran. Immediately after sunset, the fast is broken, prayers are said, and then Mr. Rahaman goes to the mosque. It is a spiritual retreat for 29 days. During the regular night prayers, many devotees do 17 rakaats and 20 extra. On the 27th day of Ramadan, an all-night vigil is kept, prayers are said, and some read the Koran. This is probably the first night on which the prophet received his revelation, and is "a fertile time for spiritual devotions."

When the new moon is seen again, Eid ul Fitr begins just as Ramadan ends. A special service is held in the mosque in the morning at about 10 a.m. Friends visit one another's homes, and charity is offered to poor relatives.

Bakra Eid

Bakra Eid, or Eid ul Azha, is a feast to commemorate the sacrifice of Ishmael (Isaac) by Abraham. This sacrifice is repeated in Mecca annually, and is the reason for the *Haj*. In San Fernando, Muslims attend the mosques, do two *rakaats*, say prayers, and listen to a sermon dealing with sacrifice. Muslims who can afford it perform animal sacrifice: a goat or a sheep for a single person, while a group of seven might sacrifice a bullock. The butcher must be Muslim, and he must cut the animal's neck, letting the blood run out.

Haj

An increasing number of Trinidadian Muslims have been going on *Haj* over the last three or four years, taking advantage of group travel. It is compulsory if you have the financial means, and about 20 went in 1963. Haji Shafiq Rahaman and Haji Ahmed Khan are both associated with the ASJA mosque. *Hajis* are very highly regarded among Muslims.

Mosque Elections

The imam is elected by the congregation or by the Executive of the Board. The ASJA Mosque Board is elected every January, and consists of Faizool Rahaman (president), Dr. Abdool (vice president), Hamza Mohammed (sec.), Roy Mohammed (asst. sec), Noor Shah (treas.), Haji Shafiq Rahaman (trustee), and Haji Ahmed Khan (trustee). Mr. Jaleel now takes a back seat. The congregation gives an annual voluntary donation toward the running costs of the mosque; businessmen contribute most of the money.

Differences between ASJA and TML

What are the differences between the Muslim League and the ASJA mosques? There are doctrinal differences. The Muslim League mosque was the original one, though the two were united until 1910 or so. The ASJA group built their own in 1914. The Muslim League flirts with the Ahmadi doctrine and its members behave like Ahmadis. They welcome and promote lecturers, but they make no open declarations. The crucial difference between the Muslim League and ASJA is that the former do 8 not 20 additional *rakaats*; and they believe in the finality of Mohammed's prophethood. In San Fernando, the doctrinal difference is not so prominent; they still regard each other as Muslims and they attend one another's functions, such as marriages.

The oldest Muslim body in Trinidad is the Tackvyatul, under Ahmadi control—hence the formation of the orthodox ASJA. In about 1940, Tackvyatul merged with the Tabligul Islamic Association, which had been spearheaded by ASJA teaching, but was not completely orthodox. Nowadays, support for the three groups splits as follows: ASJA 44 percent, Trinidad Muslim League 28 percent, and Tackvyatul 28 percent. There are about 40 ASJA mosques and 20 associated with the other two.[21]

Politics

Politically, "Muslims are a minority within a minority," and it is not safe to show your hand. Major politicians are affiliated as follows: Kamaluddin Mohammed (Tackvyatul, San Juan and PNM), Saied Mohammed (ASJA, Port of Spain and PNM), Nazim Muradali (TML and DLP). Muradali lost to Gerard Montano (PNM) in San Fernando East in 1961. There is some truth that Muslims tend to support the PNM, but not completely so. They are more open to the PNM because they are a smaller group than the Hindus and their religious teaching is interracial. Importantly, most Muslims think of religion before race; the head of the Islamic Missionary Guild is a black Muslim.

Mr. Rahaman expressed admiration for Bisram Gopie and the Seva Sangh, the Hindu Temple and its activities. He did not introduce me to his wife, though she teaches at his school, and seems to be fairly emancipated.

Friday, July 31

Colin: *Miss Anna Mahase, Principal St. Augustine High School, Tunapuna*

Gillian and I went together to interview Miss Anna Mahase,[22] Principal of St. Augustine High School near Tunapuna. She took her Cambridge Higher School Certificate at St. Augustine High School aged 16, and studied Chemistry at Mount Allison University, New Brunswick, Canada—the university of the United Church of Canada. She returned to Trinidad on graduation, and had been on the staff for a relatively short time before becoming principal in 1961.

Miss Mahase's mother had been the first East Indian woman teacher in Trinidad. But both sets of grandparents were immigrants from India. Her maternal grandfather came as a Presbyterian minister, while her paternal grandfather, who traveled on the same ship, was a pundit. Miss Mahase's mother, therefore, was born a Presbyterian, while her father was a Brahmin, who converted to Christianity, went to Presbyterian schools and Teacher Training College, and eventually took a BEd. externally. He became a primary school headmaster at Guaico, near Sangre Grande, where he served for 40 years. Her mother studied on her own, and later attended Naparima Training College.

Miss Mahase's parents had seven children, all of whom were taught by their mother to play the piano. The Christian influence in their home was a great advantage: "only those who accepted mission teaching made anything of their lives." All seven went to university as a result of the teaching of the Canadian Mission, and the mature outlook of her father, which she ascribes to Christianity.

I asked what Miss Mahase thought about being an East Indian. "I never really think about this—unless the history of the family is involved, or a direct question is asked." Miss Mahase never thinks about race or ethnicity when disciplining children. "I am concerned with the person as an individual."

However, Miss Mahase is conscious of being Indian, and is a member of the Indian Club of Port of Spain. It has very few individual members, mostly husbands and wives (including white wives). If you see a black person there, he or she is usually a visitor. The Trinidad Country Club is still largely white.

When Miss Mahase goes abroad—especially to the United States—she is referred to as "the young lady who comes from India." She does not wear a sari, though she was advised to take one to Canada when she went as a student; she still has no idea how to put one on. Even now, she describes the sari as "not normally my dress." Saris are for Hindus, and Miss Mahase "can see nothing in looking to India." Presbyterians are western in outlook, and very much influenced by Canada, though racially East Indian. "It leaves us where we should be—a western person." Presbyterian Indians have more in common with any Christian than with Hindus.

Most of her friends are East Indian Christians, for example, colleagues at the school. "Deep down inside, East Indian Hindus are proud of non-Hindu Indians who get to the top." She added that she felt that her own appointment had been "premature." The previous principals had been white and Canadian. About 65 percent of the St. Augustine girls are East Indian; 35 percent are black. But, when scholarships are awarded by Miss Mahase, she never notices to which race they are given. There has never been report of racial conflict in the school.

Sunday, August 2

Colin and Gillian: *Arya Samaj Wedding, Chaguanas*

We attended an Arya Samaj wedding at Montrose, Chaguanas—Gladys Maharaj of Longdonville was marrying Ramkumar Singh. Both are teachers at the Montrose Vedic School. After the *barat* had gone to Longdonville to collect the bride and bring her back to Montrose, the ceremony began with the agwani (meeting-up) of the fathers of the bride and groom outside the Arya Samaji temple. During the agwani, the fathers garlanded one another with flowers. The *maro* was made up of a canopy with streamers, the backcloth to which was a candlewick bedspread decorated with a paper sun in which was set the Om sign.

Bride and groom were beautifully dressed—the bride in a white sari embroidered with gold silk, the groom in a white *dhoti* and cream silk *kurta* with a gold collar, pink turban, and pink-and-white striped scarf. The bride's father was more modestly turned out in a shirt and pair of white trousers. The bride (seated to the right of the groom), her father, and the groom washed their hands and twice drank water from them. The kanya dan was performed with the groom's right hand under the bride's with the father's on top. Both the father and the groom repeated prayers spoken by Pundit Sitaram from San Juan.

The *vedi* consisted of three tiers of interlocking triangles, the lower two decorated with green and ochre rice, and the top tier containing deep within it the fire. During the *hawan*, the couple fed the fire, previously lit by the groom, with pitch pine dipped in *ghee*, while a Vedic *mantra* was said. Additional *ghee* was poured onto the fire by the groom from the pundit's Old Oak rum bottle. *Bhajans* were sung accompanied by a harmonium. The bride fed the fire, in unison with the groom, by spooning wood and raisins over it—as in a Sanathanist sandhya puja. The bride and groom then stood facing one another wearing garlands, announced their willing acceptance of one another, and garlanded each other with the flowers they had been wearing.

The bridegroom's acceptance of the bride's hand (pani grahan) involved the girl placing her right hand between the hands of the groom (right over left). The ceremony moved rapidly to the signing of the marriage register, followed by the mixing of *lawa,* brought from the parental homes in large paper bags, and the throwing of three handfuls on the fire by both the bride and groom. *Mantras* were recited as the groom placed a wedding ring on the bride's finger.

The reception was held in the groom's home. During the reception we had a conversation with Mrs. Rajkumari Maharaj, widow of Pundit Suruj Balli of San Fernando, and later, with their son, Mahendranath. Suruj Balli had been a Sanathanist pundit before he joined the Arya Samaj. Mrs. Maharaj gave us a list of Arya Samaj pundits: Sita Ram (Brahmin) from San Juan, Vishnudath (Brahmin) from Diamond, Latchmidath (Chattri) from Chaguanas, Rhadigosain (Brahmin) from Arima, and Baldeopersad (Sudra) from Débé.

As the majority of Arya Samaj members can perform their own *hawan*, there is not so great a need for pundits. Any male Arya Samaji can take the *janeo*, but, if they do, they must not eat meat or drink alcohol. Among Arya Samajis, caste intermarriage is easier than among Sanathanists: Mrs. Maharaj was a Chattri, her husband, a Brahmin. Mr. G. B. Lal is Mahendranath's godfather.

Mahendranath told us that he goes out on dates, though he has been turned down by girls from strict Hindu homes. Recently five couples went from San Fernando to a dance at the Hilton Hotel in Port of Spain: two couples were married (all Presbyterian Indian), two couples were engaged (one pair Muslim; one pair Indian/ Presbyterian and Portuguese Creole); and the last pair—including Mahendranath himself—was Indian (Arya Samaj and Presbyterian). But Mahendranath's mother will not let her daughters go out on dates, because, if they do, girls lose their reputation in the eyes of strict Hindu parents.

Monday, August 10

Gillian: *Capildeo Maharaj*

We visited the home of Mahadeo Pundit and his family in California, and then went with them to see Capildeo Maharaj, Mahadeo's younger brother. Capildeo once practiced as a pundit, and had a sizeable following of his own. Now he is an English master at Presentation College, and is the school's first Hindu teacher. He also owns a groundnut estate at Arena, beyond Freeport, where the family property has been split between the brothers.

Capildeo is full of paradoxes, having some of the sophistication of the town teacher and much of the earthiness of the country estate-owner. As a teacher and farmer, he is comfortably off, yet he lives on his land in a spartan, little wooden house. In his garden the Trinidad national flag and *jhandi* fly side by side. He eats no flesh, yet smokes heavily—always putting his cigarettes and cigarette ash well out of sight when Mahadeo is around.

Capildeo is modern, unconventional, yet a devout Hindu, and he has a crude quick wit. For example, his remark to a finicky aunt who refused to eat with her fingers, because it is insanitary. "What is wrong with her hand when she has washed it? If she doesn't think it's clean, then she should chop it off after she has been to the lavatory!"

Capildeo was married "under the bamboo." He referred warmly to the wedding day, and is unusual in speaking openly of his fondness for his wife. Yet they look a typically Hindu couple in so many ways. She wore the *oronhi*, that is to say she had a chiffon scarf over her shoulders (perhaps that, too, was for Mahadeo's benefit). She stays at home and cleans like the good Hindu wife. But they have only two children (aged six and five).

The estate was smallish and set in gently rolling, sandy country—loose sandy soil is good for groundnuts. Here they grow cocoa, coffee, and some tobacco, though at this time of the year the main crop is groundnuts. It is the only groundnut estate in Trinidad, and

Capildeo sells to Deabi Persad Maharaj in San Fernando. We saw cocoa and coffee beans in a drying house, and a huge *crapaud* (toad) in the long, wet grass. There are wonderful views from the estate toward the Northern Range of Trinidad and the coastal mountains of Venezuela.

The number of Maharaj relatives in the Arena area is astonishing, and includes Pundit Mahadeo's wife's family as well as his own. Pundit Mahadeo is a fund of information on family connections, as befits someone who has conducted many marriages. Mahadeo's family is related to the young man from La Romaine who married Vedwatie Maharaj (Ramnath's daughter) on April 12. He also told us that Dharam Pundit's wife used to sell in the market in all her jewellery. She is a sister of Gobardhan Jerrybandhan's late mother and Miss Harilal's mother (all members of the Ahir caste, I think). One especially interesting connection was that Bisram Gopie's mother was Hansar Ramsamooj's father's sister.

Tuesday, August 11

Colin

We attended the farewell for Rosalind Ramsamooj at her parents' home in Corinth Settlement.

Alloy Lequay, businessman, on sports clubs

Earlier in the day I had had an interview with Alloy Lequay, an Indo-Chinese acquaintance, who is branch manager of the West Indies Insurance Company. He had been the unsuccessful DLP candidate in San Fernando West at the 1961 election. He commented, "I am more interested in sport than in politics. In sport you make friends and in politics, enemies."

Oriental Cricket Club is an East Indian team; it is not a sports or social club. The team is formed every year. It was set up more than 50 years ago, but has no ground or clubhouse of its own. Gopaul has donated a ground on Marabella Lands, north of the Vista Bella River, but probably no deed has been given—it's just a technique for developing the lands.

Oxford is the only independent club to play competitively all the year round. It was East Indian by accident, not design. It was founded in the Gomez Street area, between Paradise Pasture and Rushworth Street in 1944, and met originally under Mr. Jaggernauth's house. The core group was East Indian and Presbyterian. Mr. Lequay was the only non-East Indian in the founding group—but he lived in the area and was interested in table tennis. In 1960 the clubhouse was built with a dance floor, members' bar, and courts for volleyball and badminton. Skinner Park is used for tennis, cricket, and football. It is now proposed to sell the clubhouse on Lady Hailes Road and buy land from Usine for BWI$120,000 (£24,000) to make a new sports ground at Corinth.

The president of Oxford Club is Percy Ramkerrysingh (Carib Lager), whose wife is the secretary at Naparima College. Oxford Club has a strong connection with Naparima College and actively recruits there. Leading members of the club now include Dr. Ramnarine Jaggernauth, Alloy Lequay, and Edwin Fyzul Rahaman. The club has lady members.

Wednesday, August 12

Colin: *Mr. T. M. Rahaman on Muslim rites and practices*

Mr. T. M. Rahaman provided me with a long interview on Muslim rites, the mosques, and individual leaders. He told me that *julaha* is used in a derogatory sense to mean "low caste." Patans are Khans. But in the present context in Trinidad, Muslim "caste" affiliation has no meaning other than for 1 or 2 percent of the Muslims. Mr. Rahaman does not know which "caste" he comes from; there is no particular pride in being a Khan.

Birth and childhood

About seven days after birth, a baby's head should be shaved and its weight should be the measure of the money given to charity. Nowadays neither of these practices is common. After seven days, akika should be performed—two animals being ritually sacrificed for a baby boy, and one for a girl. It is appropriate that the poor should be fed from the animals, but not the parents. The uneaten parts of the animals should be buried. Some use *kajar*, when a child has a cold. Black dust (*surma*) is used as a beauty treatment. Indians, generally, are fond of jet beads, since it "cuts *maljeu*," or *badnazur* (evil looks). A boy has to be circumcized, the younger the better, but usually before the age of 12. There are professional circumcizers, one in Princes Town, but none in San Fernando. There is no special time for a girl to start wearing the *oronhi*. Menstruating women should not have sexual intercourse: for seven days, they should not pray or fast (during Ramadan), but can cook and serve food.

Marriage

Many more marriages are arranged than is apparent, though there are both arranged and free-choice unions. Islamic law does not recognize a marriage that does not have the consent of both parties. Engagement involves an exchange of gifts, a ring, a prayer of blessing, and the sharing of a common meal—usually at the girl's house. In some cases, a double engagement takes place, with the boy and girl making a formal visit to one another's homes. Dating varies from family to family, with urban Muslims being much more flexible than the rural ones.

About 90 percent of marriages involve proxy. *Barka-chotki* (husband's eldest brother/wife) avoidance relations are not really strong among Muslims, for whom religion and social practice are not as bound up as among Hindus. Muslims can marry cousins, but there are certain prohibited degrees. About a week after marriage (or after the honeymoon), the bride spends a short time visiting her parents. This ends when the groom spends a day at his in-laws, collects his bride, and takes her home.

Death

The head is tied and mouth closed; the eyes are closed; the hands are placed at the sides; and the toes are bound together—all before the body gets cold. Incense is burned until the funeral. A wake is held, which involves a reading from the Koran on the theme of death. The body is prepared for burial, bathed (and mouth washed out), and dressed in a long white cotton gown or a suit. Burial takes place with the body orientated north to south and the head turned toward Mecca.

The preferred place for the funeral service is a mosque, but it can be held at the graveside or at the home of the dead person. Prayers are performed standing; then the body is interred, the grave covered, and the final prayer is said. There are no wreaths. Three days later, *niaz* (prayers) are said for the dead person; a reading from the Koran is given; incense is spread over lighted coals; and the poor are fed. After the 40th day, the same ritual is followed, and again on the anniversary of the death.

Mouloud Sharif

Mouloud Sharif (the term *kitab*, meaning a reading from the Koran, was the preferred term in Trinidad when Mr. Rahaman was a boy) may be celebrated on any joyous occasion, for example, on the evening before a marriage, or on the occasion of a betrothal or a circumcision (but not when a death has occurred). Mouloud Sharif may also be performed to mark the return from a journey, the conclusion of a business deal, or the passing of an exam. It has a strong social value.

The ceremony involves reading from the Koran, singing songs in praise of God and the prophet, and a talk on an Islamic topic. This may relate some interesting event in the life of the prophet. An imam normally officiates, with the help of others. Usually the reading is in Urdu, and it is sometimes translated into English. The speeches are largely in English. There is very little speaking in Urdu; more stress is put on Arabic. The reading of Arabic characters is taught in the ASJA primary school, and on Thursdays there is an adult Arabic class (not well attended) taken by the acting imam.

ASJA Mosques and others

Gool Aziz is the assistant imam, deputizing for Mr. Hosein, because he is ill. Gool Aziz wears an overall to distinguish himself from the rest of the congregation. The new acting imam, Mulvi Said, is a huckster or peddler from Marabella. Mr. Hosein speaks "beautiful and fluent Urdu," and will die as imam. He has been imam for the last 15 years. He is the uncle (mother's brother) of the two other Mr. Rahamans (Shafik and Fuzloo), and has been court interpreter and film censor. Urdu vocabulary has been kept alive through film, but there has been some loss of Urdu, because Arabic is the only sacred language for Muslims.

Islam is taught for 45 minutes every day at the ASJA primary school, and there is some instruction in Islam at ASJA College. The head, Mr. Fazul Ali, was living in England for approximately 12 years.

The Marabella mosque is ASJA, too. Its Friday service is presided over by Mr. Hashmat Ali (a vendor in the San Fernando market). The building is not used as a school. People must take the initiative, but ownership is usually vested in the ASJA committee.

One of the oldest mosques is in Victoria Village, just north of Golconda. Before this mosque was built, an annual meeting of Muslims took place in Palmiste Pasture, between Duncan Village and the sea. During indenture it was impossible for Muslims to assemble for Friday worship in the mosque at 12.30 p.m. (after midday), so the *kitab* was the binding force in the community. Songs of praise (in praise of God and the Prophet) were sung, and narratives read aloud relating incidents about the Prophet.

Local leaders and the community

Mr. Shafik Rahaman, who has been president of ASJA for about five years, was vice president long before that. Former presidents were Haji John Mohammed (now honorary president) and Haji Mohammed Ibrahim (before Haji Mohammed). Before 1951, ASJA was essentially based in Port of Spain. Between 1941 and 1951, San Fernando was associated with ASJA only because Mr. Hosein was such a live wire.

The Islamic Missionaries Guild has the objectives of keeping Muslims together and of providing information for non-Muslims. It has a radio program and it publishes leaflets. The aim is "to learn a little more; to practise what we know; and to teach what we know." There are about 2000 Muslims in San Fernando of whom possibly 500 are staunch believers.

Mr. Rahaman's father was a *hafiz* (able to recite the Koran), having studied theology in India. He was one of the founders of the mosque, and a full-time religious leader, kept by the community. For many years, Mr. Jaleel was the backbone of the ASJA community and the ASJA mosque was called the "Jaleel Mosque." Now the Rahamans play that role. Haji Shafik Rahaman did not have such wealth or influence; Fuzloo played second fiddle to his father, then to his brother. Fuzloo might not be president next year. The imam is elected by the mosque board; and the mosque board is elected at an annual general meeting. The color green is the color of Islam, hence the white and green of mosque architecture.

Food

All animals that are eaten must be ritually slaughtered. Crab is disliked by Muslims; pork is forbidden. *Parata roti* is strongly associated with Muslim Indians, though it is eaten by Hindus too.

Social mixing

According to Mr. Rahaman, slavery is used by Creoles as the explanation for everything; and the cocktail party is treated as a form of compensation. The social parties of the elite, whether Muslim, Presbyterian, Hindu, or Creole, are all the same. The Lions Club is the latest exclusive group. Ralph Narmi and Nazim Muradali, both of the TML mosque, are Muslims who are highly sociable. Oxford Club is the meeting place.

The Rahaman brothers will not mix socially with the other groups, except out of politeness. They tend to mix with other Muslims, though Shafik Rahaman was formerly chairman of the Southern Chamber of Commerce.

Preparations for the Seukeran wedding

Later that evening we went to the house of Mr. and Mrs. Carl Seukeran. Carl Seukeran takes pride in his caste. He does not bow to a pundit, because he is the pundit's equal. He comes from a long line of pundits on both sides of his family, and his father was a graduate of Benares University.[23] Brought up steeped in Hindu tradition, he converted to get a job as teacher with the Presbyterian Board, and others converted through him. He experienced a change in attitude with the formation of the Maha Sabha, and his reconversion took place when his brother, Lionel Frank, got him the job as head of the Hindu School in Siparia (where he is required to teach Hinduism).

Mrs. Seukeran commented that Carl has a foot in two camps, and we noted that he deliberated before saying he is a Hindu. Yet he calls in a pundit from Tableland to do *rōt* (Hanuman puja) for him. Carl has the same voice and mannerisms as Frank (Lionel Frank Seukeran, the legislator). Lionel Frank is clearly regarded by his brother and sister-in-law as a devout Hindu. Carl referred to his own impulsive decision to give up eating meat, though it is an inconvenience for his family; his wife would rather he gave up drink! She stressed the importance of being a good Hindu or a good Christian.

The Seukeran second son, according to his parents, is marrying "a Portuguese"—presumably of Madeiran origin—though a friend of ours claims she is "a black girl." A Muslim woman who is visiting chipped in: "I just don't know where they come from. Not even the animals does have kinky hair."

The Seukerans' eldest son married a Muslim when he was in England, and this has helped to break down Carl's resistance to a mixed marriage here in Trinidad—the bride is Roman Catholic. Mrs. Seukeran is racially more tolerant (which must have helped), but even she would not have liked a black Creole. Carl does not seem to think much store can be set by his son's promise that children born of his marriage will be brought up Catholic.

The Seukerans take enormous pride in their only daughter. All their children have been indulged, she even more so. Rather than let her mix, Carl takes her with him to school in Siparia, though, educationally, it would be better for her to go to school in San Fernando. She is afraid to play in the street, because they live in a predominantly black area and she gets pelted. Her mother says she would not let her go abroad to study because she would not be able to keep an eye on her there. In contrast, the Muslim visitor, who says that her 18-year old daughter has not been dated, does not seem to care what will happen to her if she goes to do nursing in the United Kingdom. She says she wants a white son-in-law.

For the wedding, they will invite DLP workers from the Naparima Ward (Lionel Frank's constituency), some friends, and family. The reception will be at the Oxford Club, and there will be Chinese food and a dance, all of which is to be financed by Carl. The Seukeran brothers are club members. The bride lives on Harris Street.

Carl wants all the wedding preliminaries, leading up to the Catholic ceremony, to be according to Hindu tradition. On the Thursday, Friday, and Saturday preceding the wedding, there will be Hindu songs and dances, with a big dinner, including goat, on the Saturday afternoon. His wife, however, does not want the group of dancers who perform suggestive and vulgar movements. She does not mind Hindi songs, which she knows are vulgar, because she cannot understand them. What she would really prefer is to hold Presbyterian house prayers. All their neighbors will be invited to the wedding, and there will be alcohol at the reception.

Mrs. Seukeran knew nothing about Hinduism until she married, and she could not cook *roti*. Her father was a Presbyterian minister. But now she believes that "is all de same God." However, she does not like the way that Hindus refer to gods in the plural. She told us about the disgust she felt for Pundit Ragoonanan of Palmyra, the husband of Carl's sister, who jumped from window to window in a Hanuman trance. "His face became just like a monkey." She ran away in fright. He drank boiling milk, but it did not burn his lips. Afterward he did not know what had happened.

Mrs. Seukeran's father was a Maharaj, proud of his caste and a strong believer in the caste system. Mrs. Seukeran, who actually married within caste, has no time for that. When her widowed father remarried, it was to a woman from Grenada. She was not a Maharaj; in fact, she did not make *roti,* so un-Indian was she. Mrs. Seukeran did not know that only Brahmins could become pundits; she thought *saddhus* could be pundits too.

Thursday, August 13

Gillian: *Hansar and Bram*

Hansar told us that Rosalind raised the money to go to England from Rama Maharaj—through Mr. Gaya. But it came from a Brahmin; as did Hansar's money from the Stri Sevak Sabha—ultimately from Deabi Persad Maharaj. Brahmins act as bankers to the Hindu community! Hansar is thrilled because Rosalind bent down and touched his feet at the airport before she left Trinidad; this meant a great deal to him.

Bram was outraged at our description of the religious "callaloo" of Carl Seukeran's son's wedding preparations; it is, however, not unheard-of for Indians to construct a *maro* while holding a Christian wedding for their children. Christians frequently look for marriage within caste, and sometimes even rub saffron on their children before the ceremony.[24]

Bram and Hansar agree that there is an unseen barrier around the Hindus, which puts off many Christian Indians (such as Peter Dubé) who are exploring the idea of being reabsorbed into Hinduism. Mr. Bhattacharya's Gita class, for example, is perhaps 30 percent Christian. The Arya Samaj was a great success when it was first introduced into Trinidad about 20 years ago.[25] They were the first to bring outside lecturers, and they captured many intelligent people. But, ultimately, according to Hansar, they became "too critical" of others and of themselves. Now they are simply hanging on to the few members they have.

Out of respect for his father, Hansar would never speak to his mother in his father's presence. She would always walk behind her husband. Mr. and Mrs. Ramsamooj would probably do the same, but they usually go everywhere in their VW van.

Bram commented Christian Indian weddings have many similarities with Hindu equivalents. All the neighbors are invited. A tent, constructed of bamboo-and-galvanized metal sheets, is frequently erected outside the girl's house for the reception. The food consists of meat, *dalpuri* and *roti*, or sandwiches and ice cream, or Chinese food. At the reception, there are speeches, a toast, and of course alcohol. Bram insists that "ignorance" was the basic cause of the drift from Hinduism to Christianity.

A white man at Texaco asked Bram for a copy of the Karma Sutra. Bram had not heard of it, and so asked Mr. Bhattacharya, who recommended it as a good depiction of life and love. Bram found a paperback copy with the picture of a half-naked woman on the cover. It was not on display in Fogarty's in San Fernando but kept at the rear of the bookshop. The text was interspersed with nude pictures that horrified Bram. His brother Hari asked him to take it out of the house, away from the reach of the children. According to Bram, it was "pure sexology."

Bisram Gopie was hoping to use SSDM support at the last election (1961), if he had contested the Naparima seat for the PNM against Seukeran, which he did not. But he had no cooperation from Hansar.

Friday, August 14

Colin: *Mr. Chinnibas, Seunarine leader*

I had an interview in Débé with Mr. Chinnibas, one of the leaders of the Seunarine sect in the south of Trinidad. The majority of Seunarine are very poor. He showed me the book, Guru Anyas, open at a picture of a temple, and mentioned the "idea of the church as one foundation." The Seunarine Dharma Sabha of Trinidad and Tobago was incorporated in 1944. The Dharmacharya is Janglie Ram of St. Charles Village, who has been the spiritual leader since its foundation. He came from India as a child. Chinnibas became an assistant *mahant* in 1913, and was elected a full *mahant* in 1925. Other leading figures in the sect include Rampaul, the president and Ramjattan (of Diamond). There has been a huge falling-out over Ramjattan's alleged appropriation of the educational fund, and a court case is pending to decide the split.

With Chinnibas I went over the Seunarine ceremony I had previously witnessed. The *dewan* (district secretary) of the Seunarine Dharma Saba had conducted the *puja*. There were three senior *mahants*: the *dewan*, who can play the drums; Janglie Ram, who is also a very good drummer; and the man who went with me, Jun Numgra, from Monkey Town. There was one Maharaj present, but those Brahmins who are Seunarine rarely admit to their high caste. We talked about the various stages of the Seunarine ceremony: likanimul, the writer of books; kowsil, observing mistakes and making judgments; charidar, gathering flowers and making garlands (*mala*); and bhandari, preparing *persad*.

Rampaul came over as comparatively rich. His children have been educated in India. I imagine that Rampaul and Janglie Ram are Sudra. Chinnibas said, "Caste is changeable; you are called by the work you do. Brahmins try to squeeze down the other people."

I asked about the Seunarine Samaj in Débé, and gathered that it is not often held in the temple. Night readings of Guru Anyas are held. The present Seunarine temple was built by the father of Dharam Nanan, who has a big house by its side. Over 60 years ago, the first Seunarine temple, which stood opposite the present one, was erected by Mooti Das. Dharam Nanan married Mooti Das's granddaughter.

Chinnibas reads the Ramayana and comments, "we are all Sanathanists." But he holds a Kali Mai ceremony annually in the back of his house to the east of the Débé Main Road. Chinnibas claimed that 60 percent of Hindus in Trinidad are Maha Sabha, 12 percent Kabir Panthi, 6 to 8 percent Arya Samaj and 20 percent Seunarine.[26] Many Kabir Panthis go to Seunarine functions. Chinnibas has not taken the *janeo*, though he argues that "anyone can take it." Fifty years ago, he used to eat meat, but now he is vegetarian and never touches alcohol.

Gillian: *Peter Dubé's Niaz*

We attended a short Muslim prayer meeting, or Niaz, at the home of Peter Dubé. His wife, formerly Muslim, is now an active nonbaptized Presbyterian. An old, knowledgeable Muslim in his seventies, who came from India in 1912, called in to say prayers. This is an annual thanksgiving ceremony in the Dubé household, held on Peter and Seeta's Muslim wedding anniversary.

A small fire was burning in an earthenware pot on a table. The only decoration was a large vase of anthurium lilies and a few other flowers. A dish of *halwa* (sweetmeat), a glass of water, a plate of curried chicken, *channa,* and *roti* stood on the table.

The Muslim drew his chair up to the table and began uttering some passages from the Koran, while Peter sprinkled lumps of camphor on the fire. The elderly man had never seen camphor used in a Muslim ceremony before, though camphor is used in Hindu rituals, as Peter was doing, to keep the fire burning without smoke. Near the end of the Koran passages, not a word of which was comprehensible to us, nor probably to Peter and Seeta, the celebrant asked the purpose of the prayers. They replied that they simply wanted to give thanks, not to promise anything. Prayers were offered in Urdu, and the ceremony was over.

The pictures in their house were unusual. There were two pictures of Christ, one above the other, and a colored photo of Peter and Seeta on their wedding day, with Peter wearing a turban and the couple standing next to a white wedding cake. The house also has a copy of the Koran in English.

The Dubé caste, according to Peter, is supposed to know two Vedas[27] by heart. Peter says that a lower caste Presbyterian would be elated to be able to associate with a high-caste man. "In-caste" marriage is important for Presbyterians.

Colin: *Indian High Commissioner's Farewell*

We attended the Farewell Reception in San Fernando for Mr. Nair, the Indian High Commissioner to Trinidad and Tobago, organized by the Gandhi Seva Sangh. Mr. Gopie (chairman of the Seva Sangh) acted as MC, and was assisted by Dr. Avatar (vice chairman) and Baldeo Maharaj (secretary). Excluding us, there were 40 local participants, of whom 33 were East Indians (31 Hindu and two Muslim—the Rahamans), including the leader of the opposition (Mr. Stephen Maharaj[28]), and seven Creoles, including the member of the House of Representatives for San Fernando East (Gerard Montano), the town clerk, and an alderman.

Saturday, August 15

Gillian: *Indian Independence Day, Golconda*

We accepted an invitation to attend celebrations of Independence Day at Golconda, naively imagining it was a premature celebration of Trinidad's independence (August 31)—not India's. Hansar made no attempt to clarify matters. We discovered that the national newspapers were full of congratulatory adverts from leading businessmen—Gopaul, Rahamut, and so on.

Presbyterian wedding, Esperance

Later, we attended a Presbyterian wedding in Esperance, just south of San Fernando. The bride and groom were both Presbyterian, but the boy's parents and the girl's father were Hindu. Influenced by the bride, her mother and brother had recently converted to Christianity. The guests assembled in a bamboo-and-galvanized tent erected in front of the house. The men and children congregated there, while the women thronged up the steps and into the tiny house to get a peep at the bride, who was being packed into her elaborate dress by a seamstress.

At about 2.15 p.m., the groom's car passed the house with blaring horn, followed by a procession of his guests. Almost half an hour later, the bride left for the church with her mother in the bridal car, and we all followed. The bride was wearing a white wedding dress, made of satin and smothered in lace, appliqués, and bows—elaborate but awkward for her to manage in addition to her bouquet of plastic flowers. She had three bridesmaids in attendance, wearing short green satin dresses. In comparison, the groom looked somber in a gray suit and tie.

The starkness of the Presbyterian church was matched by the simple ceremony, though there was a forceful sermon by Rev. Baldeo, stressing the triangle of the Christian marriage involving God, the bride, and the groom, and the need for love and rejoicing. Procreation was not stressed; in fact it was barely mentioned. There was a noticeable lack of the village women in the *oronhi*—they had come to the house but not to the church, though they were around for the reception afterward. The car, containing the newly married couple, set off for a drive around the country lanes leading through the cane fields for a while, before returning home.

As with most Christian weddings, the reception was, from our point of view, the most interesting part. It was held in a small room in the bride's house. A long table was set, with the bride and groom at the head, the bridesmaids and the guests of honor (including us) arranged around. Rev. Baldeo acted as MC. Toasts were offered first to the bride, then to the groom, and finally the bridesmaids. All the speeches were short and Christian in tone. Whenever a speaker made a good point, there were enthusiastic cries of "Aymen, Aymen!" Prayers were offered for the bride and groom (whose health was drunk in Cydrax), and the doxology was sung.

The "sticking" of the cake involved the bride and groom each holding a knife and cutting either side of a thick slice. As they cut, a quavering version of "And the wedding bells were ringing" was sung until the couple's knives met. After this we had grace led by Mr. Coleman, the headmaster of the school in which the bride teaches.

Knives and forks were laid, but the bride and groom ate with spoons. So did almost everyone except Mr. and Mrs. Coleman, who used knives and forks. The meal was *parata roti*, goat, *dal*, salad, rice and curried peas, with water to drink. We were struck by the delightful incongruity of western-style bridesmaids in all their finery eating with their fingers. But it was a sincere and simple function, and there was no music blaring. As far as I know, nobody was offered any wedding cake to eat. The cake was traditional, with icing flowers and bells, and a white bride and groom on top. There were two side cakes, one of which was heart-shaped.

Presbyterian engagement, San Fernando

On Saturday evening we attended a Presbyterian engagement. Lynette Ballack from Marabella was getting engaged to Saul Gajhadar from Rio Claro. A vase of anthuriums and a closed Bible had been placed on a long table, covered with a plain white cloth and decorated around the edges with lots of fern. Friends gathered around the table, and spilled out onto the gallery and steps. At the top of the table sat the betrothed couple and the pastor, Rev. Mootoo from Marabella.

Prayers were said and a psalm, "The Lord is my Shepherd." was sung, which was followed by a Bible reading (the story of Rebecca). Love was stressed as the essence of marriage, as was the need to pray to God in the choice of marriage partner. Ultimately, the choice of a partner is God's (Hindus would argue that God chooses through the horoscopes of the couple, via the *patra.*)

The pastor blessed the ring on the Bible, and the couple's hands were held over the sacred book for the blessing. Lynette's brother sang "The Young Ones," while the couple "stuck" the cake, followed by another love song while a married couple "unstuck" it.

Rev. Mootoo left at this point. Then the fête began with a good meal and drinks (alcoholic if wished), and was to go on into the night, with dancing upstairs in the house, and talking down below. The religious part of the ceremony had lasted 30 minutes instead of the usual 60, because Rev. Mootoo had another engagement. The gathering in Marabella was predominantly Indian, though we noticed two black friends on the gallery.

Sunday, August 16

Colin: *Dan van Eendenburg, the Gita at Point Fortin*

We went to Point Fortin to visit the van Eendenburgs, who live close to the Shell refinery where Dan works. Dan told us he had been invited by a Mr. Mathura, on behalf of Bramachari Hari Ram, to read Dr. Radhakrishnan's translation of the Gita to a Sandhya Group at Fanny Village, near Point Fortin. Recently Dan had also been invited to Gasparillo, north of San Fernando, by the Divine Life Society, but he is cagey about this. He will not usually give explanations to people; he insists that they go to their pundits for that.

Dan has a detailed knowledge of the philosophical intricacies of Hinduism. It is clear that this bears little relationship to Hinduism as practiced in Trinidad. In Dan's view, Bramachari is sincere, but his teachings of meditation and yoga are unsafe. He referred to the problems associated with a class at Sooknanan Maharaj's *mandir* at Vistabella. It seems as though the Divine Life Society will use it as a base in San Fernando. The power of the Bramachari lies in the fact that he is a Brahmin, and can give gurudiksha, which reinforces his leadership role. But he is getting some opposition from the Maha

Sabha because he is teaching yoga without experience, and is probably also competing with the pundits.

Monday, August 17

Gillian: *Duffy Mohammed on Naparima Club*

Duffy Mohammed spent much of the afternoon telling us about Naparima Club between the two world wars when he was growing up in San Fernando. At that time, the leading lights of this elite club were virtually all white: Rostant, De Brehmler, Dr. Krogh, Major Rushford, Mrs. Red Hobson and her husband, Jackson Lewis (Manager of Huggins), Whittott (Usine), Judge, Rodney Jack, George Farah (furniture), Farfan, and Miss Maclean. Sir Henry Pierre[29] got in later. Jackson Lewis still owns a great deal of property, including the arcade and Y. de Limas (jewelers), and remains a member. However, even before World War II, there were black club members such as Archbald.

One of the current social tops in San Fernando, who can pull a lot of strings, is Roy Dopson. Now a major figure in Naparima Club, he is part Indian, part white, and his father was respected. His mother or grandmother was Indian (Madrasi). His brother works in the town hall as a clerk. Roy was a right-hand man for Roy Joseph, the Syrian, in his days as a leading politician. Allegedly, he does not like Indians at all. Stella, Roy's wife, has some black ancestry, but Duffy describes them as people who "will do anything to be white. They move in top society with the Camps and the Dieffenthallers, and their family marry Barbadian whites for preference."

Colin

Duffy's real name is Claudius Mohammed. His birth father refused five acres of good oil land from Duffy's Muslim stepfather, largely because he was a Muslim. Duffy's birth father said that his great-grandfather was a Kahar (cultivator and fisherman caste), who had come from India with Rev. Morton. Duffy's birth father was a Presbyterian, like Duffy, but he used to read the Gita (he could read both Hindi and Sanskrit) and had a *tulsi* tree in his yard. Duffy thinks that caste is very important among Presbyterians.

Duffy ended his visit with the observation: "Indians are so funny: they tell you something, and deep down they mean something else."

Tuesday, August 18

Colin from Gillian's notes: *Dr. Rudranath Capildeo, leader of the DLP*

We had an interview with Dr. Rudranath Capildeo,[30] the leader of the DLP, at 65 Carlos St, Woodbrook, Port of Spain. He explained that circumstances had forced him into politics—the narrow victory of Williams in 1956, with 36 percent of the votes cast. There were two critical PNM seats: Tunapuna with a 179 majority and St. Joseph with 109. The 1956 election results are set out in the *Trinidad and Tobago Yearbook* for 1959, to which Capildeo referred.

Enter a Miss Gibbs on the urgent matter of her "very champion dog," whose entry into Trinidad from Venezuela had been prohibited. She is off to see Dr. Eric; she wants them to make an exception of her dog. Why come to Capildeo for that, we ask? But there seems to be a lot of money involved. Capildeo is beset with problems—"finding homes for dogs, and husbands for young women!"

Capildeo returned to his electoral analysis. In 1956, Tunapuna and St. Joseph were won through squabbles. In that election there was no strong racial feeling at all; there was always an undercurrent, but it was essentially a fringe disturbance. The Colonial Office gave Williams five seats.[31] "Black arrogance and bombast went to his head. God did not send Moses, but he sent Williams." From then on, black arrogance was felt on the buses, in rum shops and on the streets, and was directed at "coolies."

This was the atmosphere when Capildeo returned to Trinidad in 1958. He had been lecturing at University College, London, but had come to Trinidad to settle his mother's estate. He took a year's leave of absence from his university post, and practiced law for a year in Trinidad, having previously qualified as a barrister in the United Kingdom. At this point Capildeo had intended returning to England, but the Polytechnic was started in Port of Spain, and the editor of the *Trinidad Guardian* agitated for him to stay and become its head. So Capildeo resigned from his university post. But the Polytechnic job was nothing more than a PNM political gambit, and it still has no permanent director.

Capildeo feels that he is being driven to the wall. The DLP leaders want him to get out, but if he left, they would be blamed by the rank and file for having a party with no leader. Then the leaders would set about damaging his reputation. Capildeo's mistake was to have said that he would fight the 1961 election; he should have got out then.

But the leadership convention reported that 23 out of 24 divisions wanted him as leader. After the convention he was told to leave his government job at once, so he cleared his desk and left the office.

Bhadase Maraj has been behind the move to destroy Capildeo's reputation. Bhadase had resigned as party leader in 1959 because the rank and file did not want him; they wanted someone to take the lead against black pressure. Bhadase resigned quickly, otherwise he would have been turned out.

In 1958 during the Federal elections, it was alleged that Learie Constantine[32] (PNM) went around in Tobago, campaigning against the local candidate, A. P. T. James, and saying that "a vote for James is a vote for the coolie." According to Capildeo, "When a black gets a sniff at power, he is worse than the European."

Independence was given to Williams on a plate; no fight was necessary. Duplicity in the DLP gave Capildeo a deep shock and he experienced a complete collapse.[33]

Democratic socialism, as the policy of the DLP, is an attempt to remove race from the political arena in Trinidad and Tobago. Capildeo has been trying to press the Indian Government to say something in this context. Nehru promised that when independence came to India, help would be given to Indians at home and abroad. Capildeo has tried to use the massacre of East Indians at Wismar, in British Guiana, to get through to the Indian community. During 1964, he delayed his return to Trinidad because he wanted to see Prime Minister Shastri in London—but a howl went up here! Capildeo is in favor of the partition of British Guiana—at least as a maneuvering gambit. The campaign for partition is growing in British Guiana every day. Partition is impossible here, however, because Trinidad is too small.

Capildeo, so he claimed, could bring the British Guiana situation here overnight, though Williams is acclaimed all over the world for his handling of the situation. Williams is regarded as a socialist in London, whereas Bhadase was put forward as a reactionary Indian leader. Secretary of State for the Colonies, Ian Macleod, was allegedly "dined, wined, and womanized" when he was here before federal independence, in anticipation that he would endorse it. Capildeo noted that it was ironic that Macleod was the only politician of note to come out in favor of Profumo during the scandal to which he gave his name. Both the British and American governments are pro-black, for economic reasons—there is nothing to be gained from India.

Structurally, in Capildeo's view, the situation here is the same as in British Guiana. It all depends on the leaders at the top. The current British Guiana breakdown was accurately foreseen in 1960 in a conversation in Trinidad between Capildeo and Cheddi Jagan. And at the 1961 Trinidad election, blacks on the streets chanted, "We don't want no coolie Premier, we don't want no *roti* government."[34] According to Capildeo, the Treasury was used by the PNM for electioneering purposes.

At the Marlborough House Conference on Trinidad and Tobago's independence, Reginald Maudling (Macleod's successor as secretary of state for the Colonies) told Capildeo the Opposition had no right of veto. Independence was to be granted at all costs during 1962. There was a great deal of debate about the extent to which the constitution could be amended; but Capildeo countered that amending the constitution was the "least of our problems," and he went on to point out that, on election day, polling booths had been kept open by the cabinet beyond the stipulated closing time. Since the Marlborough House Conference, Capildeo feels that he has been simultaneously out of the political picture, but in focus.

Some Indian names, such as Ashford Sinanan's, are symptomatic of half conversion.[35] But Christian Indians have been made to feel "damn coolie" by Williams. Religion and culture are being stabilized here—for example, the movement back to Hinduism in a modified form.

Capildeo argues that the only course for the opposition is to let Williams fail economically, rather than to be blamed for agitating and precipitating failure. Capildeo is motivated by the need to preserve "the Indian race" and to break the government of Williams. Blacks have the vicarious satisfaction of knowing that "my man is on top"; the reality is that black intellectuals and the black middle classes are in power.

The only chink in the armor of the PNM is for the DLP to wait and allow the economic pressure to build up, though democratic socialism is both a "face-saver" and a reason for change. Capildeo's absence (once again lecturing at UCL) is portrayed by him as taking the edge off black animosity to Indian politics. "One big fight in a rum shop, and we'll have the British Guianese situation here." Capildeo's choice is to stay and keep quiet, but the pressure will then go off the Government, and he can't allow that. Or he can go to London, and get on with his mathematical work, which is growing in repute (which he obviously thinks gives him the best of both worlds).

Capildeo expressed himself as very surprised at black antagonism; he could not be sure what would happen if the Indians came to political power. The trouble with blacks is their emotional instability; but he is looking for multiracial support, and finds that "where blacks have benefited from civilization and breeding, as in the Sudan,[36] they are well mannered and trustworthy—very fine."

Talking finally about Hinduism in Trinidad, Capildeo contended that caste divisions will go (not the priesthood and Brahminism). "Buddha, swept caste out of India, but Brahmin priests brought it back." The idea of temple worship will grow, picking up on Christian and Jewish forms of worship. He sees the pundits as essentially parallel with the Christian priesthood; but Brahmins will continue to control the Hindu priesthood in Trinidad for a long time.

As a young man, Capildeo took the *janeo*, and he could, therefore, practice as a pundit. However, he has never engaged in ritual performed in public, though he has in private. He comes from a long line of pundits. Hinduism stresses "God within (trying to come out)," and it is the key to the Indian personality.

Thursday, August 20

Colin: *Vedic hawan in Diamond Village*

I attended a Vedic *hawan* conducted by Vishnudath[37] Pundit in Diamond Village. This consisted of Vedic prayers in Sanskrit and English and long *bhajans*. The *puja* itself was sung in unison and afterward read, verse by verse, in English by the pundit.

Vishnudath slipped off his sandals, washed his hand, and drank water from it three times. He then cleansed his shoulders, legs, back, mouth, nose, eyes, and ears. A fire was lit on a table, to the chanting of prayers. Five pitch-pine sticks dipped in *ghee* were put on the fire, together with the remains of the *ghee*. The *ghee* was contained in a metal dish, rather than in the near-ubiquitous rum bottle! Water was poured into a narrow channel running on four sides of the fire. Fuel and *ghee* were added to the fire. Food, camphor, and nuts were added to the fire as an offering. Then the fire was put out at a stroke by the pundit.

The pundit put his sandals back on, and launched into a sermon on the five great duties that man should perform daily: meditation; cleansing; respect for the elderly; feeding and caring for animals; hospitality.

Sunday, August 23

Gillian: *Gunness house party in Mayaro*

We drove across the island to spend the day on the beach in Mayaro, in a house that Dr. Robert Gunness, brother-in-law of George Sammy, has rented from Dr. Bissessar, the district medical officer. The overnighters and day visitors were a racially mixed bag: Dora Henckel (Indian) and her husband (black/German), Robert (Indian) and Jean Gunness (English), Helen and John Miller (Helen is Jean's cousin, and John, also English, is on the staff as a teacher at Texaco, Point-à-Pierre), Keith and Pearl Allahar (both Indian), a white couple from Barrackpore (English, working in the oilfields), Norman Cooper (English, working on rehabilitation for the disabled), and the two of us.

When we arrived, the party struck us as Creole in atmosphere: rum punch and beer were being handed around, and everyone was tucking into *souse* (pigs' trotters). We ate *pelau* (chicken, rice, and peas) for lunch, and there was a lot of talking.

Dora Henckel (née Mootoo) is a Christian Indian who married her husband, a black/German from Southern Africa, against her parents' wishes. The family's antagonism began to diminish when he opened a hardware shop and started to display considerable business acumen. This, of course, is a feature much admired among Indians—possibly even more so when it is found in a black. It should be added that Henckel's is the only black-owned shop on the north side of High Street in San Fernando, and that the Henckels own arguably the finest house in town, with superb views from the southern edge of Spring Vale over the shopping center and Harris Promenade.

Dora said that she had not had much contact with the Indian community—particularly the back-to-India group. Dora in no way associates herself with the east. Although she is sometimes invited to Hindu weddings, she rarely goes. She was invited to Bisram Gopie's daughter's wedding, and she is "touched and delighted" that the bride still corresponds with her. Most of Dora's friendships are with "social Indians." She also says she is fond of Ena, our landlady. Dora's sisters went to school with Ena, and Dora says that Ena is both good to Indians and liked by them.

Robert, on the other hand, regards Ena and all her friends at Promenade Club as racist and anti-Indian. I later asked Jean whether Dora and her husband belonged to the San Fernando elite, and Jean said "not really." They are not quite acceptable because their first child was born out of wedlock. Their children have married out of the community—one daughter is married to an Italian; there are sons studying in the United Kingdom; a teenage daughter is employed as a secretary in her father's business.

Jean told us that Robert was turned down for membership of Promenade Club when he applied eight years ago. So were some of his Indian friends. Robert remarked that Will Hercules and his set are determined to keep their friends in control of Promenade. Jean regards the Montanos as San Fernando's elite, not Dr. Abidh, who is rarely seen socially. Jean runs down the list of San Fernando Soroptimists: Gladys Montano (Canadian), Jean Lessey (Trinidadian, nearly white), Ena Hercules (black), Mrs. Gabe (North American), Ada Date-Camps[38] (Grenadian 'high brown'), Phyllis Nandlal (East Indian), Kimfa Winchester (Chinese/Creole), Joan Chin Aleong (Chinese), and a sister of the Montanos. Where is Ena Scott-Jack in this? Among the people we know, the following are members of Naparima Club: the Nicholsons, the Dopsons, Ada Date-Camps, the Gun Munroes, and Makhan Dubé.

Jean knows the Pointe-à-Pierre people through her cousin. Otherwise, she claims, the Texaco whites would not want to socialize with her because of her mixed-race marriage to Robert. Jean is concerned for her children because of their mixed blood. She wants a good education for them, but fears they might suffer if sent at an early age to an English boarding school. She would prefer to send them in their teens, when they will understand the whole situation better. There is some pulling and tugging going on between Jean and Robert, for example, in their approach to the upbringing of children. Robert is the typically protective Indian parent, whereas Jean leaves them pretty well to their own devices. She never urges them to eat, and, if they are hungry, they must ask for food.

Jean's family is very much under the influence of her mother-in-law, Mrs. Ivy Gunness—or simply Ma. Although Robert and Jean live separately from Ma, all sorts of decisions about the children are made by her: she says which dentist they should visit; and the children were baptized Presbyterian in accordance with Ma's wishes. George Sammy seems to have had a real struggle with Ma over his wife Myrtle, but Jean doesn't resent Ma. On the contrary, she admires her forceful character, while nevertheless thinking that, by English standards, she indulges the children.

Jean has no love of Indian food. She doesn't want her daughter to leave Trinidad before she has learnt to make *roti*, yet Jean can't do it herself. Jean knows nothing about caste. She asked us if "darkness" among Indians is associated with low caste. From the way she asked, she was clearly pondering Robert's background.

Despite having all the material comforts, there is something lacking in Jean's life. She comes across as someone who has suffered disappointment. Deep down, she seems disillusioned with Robert, who, though successful, is power- and wealth-oriented, and extremely hardworking (rather like Norman Girwar, with whom they are close friends).

Norman Girwar was a schoolmaster before his arranged marriage. He studied law in Trinidad, then went to England to become articled and take his finals (as did Ena's brother, Garvin). On his return, Norman entered the legal profession, and he now owns a big house in St. Joseph's Village plus an estate managed by his brother.

Jean and Myrtle Ma have been to Jamaica and met Pundit Tewari and his wife. Colin knew Pundit Tewari in Kingston (in 1961). George and Myrtle Sammy entertained Pundit Tewari and his wife and the Girwars on Ma's behalf the night we went to dinner, and Pundit Tewari gave us a copy of the centenary book on Indians in Trinidad.[39]

Monday, August 24

Colin: *Mr. Ramlochansingh, Headmaster Excelsior School*

Mr. Ramlochansingh was born in Siparia, educated locally, and then attended Naparima College in the early 1940s, and Naparima Training College. In the late 1950s, he took a degree at the University College of the West Indies (UCWI) in Jamaica in Engineering, Economics, and History, but in 1959 went back to teaching in his old elementary school. Soon after, he replaced the head of a private school who had won a government scholarship to UCWI. The school, which he now owns, was renamed Excelsior.

The paternal grandfather of Mr. Ramlochansingh came from India; but he does not know the town or region, and his own parents were baptized Presbyterians. Mr. Ramlochansingh understands some Hindi through spending holidays with his grandparents. His wife's grandfather was born in Jamaica. He feels Indian, but it is not racial. They eat Indian and western food in equal measure, and have no problem with meat, including pork. Mr. Ramlochansingh's paternal grandfather owned cocoa estates, which were then divided up. His father inherited 10 to 12 acres, but then went to work in the oilfields, before buying a dry goods store in Siparia.

Mr. Ramlochansingh does not have many close friends: some "black chaps at Texaco" and George Sammy he could count as his friends. Myrtle Sammy taught with him at Picton, Golconda. "Over the relatively long term, integration will take over in Trinidad. It is a well-mixed entity, quite a rapid development in town, but relatively slow in the country." "If one of my children wanted to marry a black,

I am not certain how happy I would be. I am not certain how lasting the match would turn out." Considerable adjustment is needed because of the diehard attitude of both races. Each race regards the other with suspicion. "Miscegenation produces a hardier people—I honestly feel this." Indians do not encourage intermarriage, but cannot always explain why they oppose it. Mr. Ramlochansingh offered me a whisky and soda, and there was a problem finding a soft drink.

Race is not a problem at Excelsior School. Mr. Ramlochansingh tries to emphasize harmony through the teaching of West Indian history from discovery to modern times. At school they have house prayers, though Mr. Ramlochansingh himself is not a regular churchgoer. Presbyterianism stands for social, economic, and educational advancement—in contrast to the experience of the Indian in Jamaica, where low-class status has accompanied culture loss.

Is there a Creole barrier? "For myself, I have no difficulty: by nature I am less constrained than the average Indian. It is easy to mix if one does not try to maintain one's own Indian image." Indians are more reserved than Creoles. Their patterns of activity and behavior and their food and drink are different. Indians do not like spontaneous, semi-bohemian parties. Blacks just drop in for fêtes and parties.

One of the biggest bugbears for Indians socially is their inability to change; it is a psychological drawback. Indian children at school watch others dance. Mr. Ramlochansingh feels that he and his wife are quite unusual in that they dance and mix, and he sends his daughter to dancing lessons, since not to know how to dance is a social hindrance. Even so, Mr. Ramlochansingh feels slightly more comfortable socially with Indians than with other groups. Presbyterianism is a passport for Indians: you can gravitate toward the Hindus or attend a cocktail party at the governor general's house.

Among Presbyterians, there remain some diehard opinions about caste: "remember the family they came from. . . . right down to Dom."[40] Possibly half of all Presbyterians would know their caste. Unlike Presbyterians and other Christians, Hindus and Muslims do not have any pull when it comes to jobs, especially in the civil service.

Indians are in the top echelons of San Fernando society—educationally, financially, and professionally. There are Indian magistrates, and the mayor is Indian. But this is not so in Port of Spain. In San Fernando there are Indians in the Chamber of Commerce and the Junior Chamber. However, the Masonic system is racially split into Indian and Creole lodges, L. Jaggernauth being the Indian Grand Master.

Mohammed Yakub Khan, leader Trinidad Muslim League

Mohammed Yakub Khan,[41] a leading figure in the Trinidad Muslim League, was born in Princes Town, and came to work in San Fernando—at Imperial Stores—in 1934. His father-in-law was Syed Abdul Aziz[42] also known as Kazi Ankalipha, the spiritual leader of all Muslims in Trinidad. Mr. Khan moved to San Fernando for the education of his children. He was a founding member of the Mosque Board, and its first president in 1937. In the same year, he became a registered marriage officer; an imam in 1953; and, later, a member of the Social Assistance Board. He is a prominent member of the Trinidad Muslim League.

Syed Abdul Aziz

As a child of about 12, Syed Abdul Aziz—Mr. Khan's father-in-law—had attended for two years the St. Joseph Islamic School or *maktab* in Port of Spain, where he learnt Arabic and Urdu. In those days, Urdu was well spoken and Muslim theology was taught. The organizer of the school was Abdool Gany[43] and the principal teacher was Babu Khan. Syed Abdul Aziz returned to Princes Town and married at the age of 16. He was the second secretary of the Islamic Guardian Association, starting in 1912. He was also East Indian National Association secretary for ten years, and organized a *panchayat* for the East Indians. When Hon. George Fitzpatrick[44] was made president of the East Indian Association in 1917, Abdul Aziz wrote in support of the abolition of the indenture system.

Mr. Khan's career; distinctions between TML and ASJA

Mr. Khan was a commercial clerk and later owned his own dry goods business. He was closely associated with Dr. Frank Mahabir[45] and Jules Mahabir,[46] a magistrate and father of Dr. Winston Mahabir. He was also involved in the East India National Congress of Couva, of which he was for a time vice president.

Mr. Khan set out for me the principal differences between the Trinidad Muslim League and Anjuman Sumat Al Jamaat Association:

- In the TML women have equal rights
- Women participate in the mosque congregation and at meetings
- Mixed bazaars are held
- No screen between the sexes is required in the congregation
- TML read 8 rakaats; ASJA, 20
- The Tazim or welcome song is not done by the TML

- ASJA believe that Christ will come again; ASJA stresses that Mohammed was the last prophet
- ASJA believe that prayers can help the dead 4, and 40, days after death
- At burial, TML places the corpse in a bag; ASJA buries the body

Mr. Khan has been first vice president or second vice president of the TML from its foundation in 1947 to the present day. Mr. Khan's brother is a *Haji*, and so are Mr. Aziz Ahmed, the president general of the TML, and his wife. The TML members will join ASJA in prayers for the dead, and so on, but some ASJA leaders would not join Mr. Khan in the TML mosque, especially those from Port of Spain. Nevertheless, Mr. Jaleel and Mr. Rahaman from San Fernando will follow Mr. Khan in prayers.

Mosque Board

The Mosque Board, which is elected annually by the congregation, provides primary- school premises free of charge. These are also used by the Muslim Youth Movement and for adult theology classes (for eight years given by Mr. Khan). On Sunday mornings, there are classes in Islam for 5- to 18-year olds. A *muezzin* calls the faithful to prayer every Friday, looks after the mosque, and teaches Urdu—there is a school period dedicated to Urdu between 3 and 4 p.m. each weekday.

Social divisions

Mr. Khan outlined the major social divisions among Muslims:

Patan—named Khan, warriors
Julahar—named Ali, who make clothes
Sayid—descendants of the Prophet, priestly (Mr. Khan's wife)
Mogul—(had to ask his wife, who didn't know the association)
Sheikh—took place of Julahar

Tuesday, August 25

Colin: *Mr. G. B. Lal, Arya Samaj leader*

Arya Samaj has interested Mr. G. B. Lal since he was a boy in British Guiana. His father, a member of the Banyia caste from Allahabad in India, was a Sanathanist, but Mr. Lal grew up under Christian influence, especially at school. His early life was spent in Essequibo and then Georgetown, the colonial capital.

Mr. Lal migrated from British Guiana to Trinidad in 1928, and worked for Trinidad Leaseholds (the British oil company that was later to be bought out by Texaco), as a staff member in the research laboratory at Pointe-à-Pierre. Then he moved into the grocery business with Canning and Co., eventually becoming office manager with Canning Bottling Co., a post from which he retired three years ago.

From 1933 to 1934, he promoted Arya Samaji missionaries from India to Trinidad, and in 1935–1936 joined the movement. He went through the purification ceremony of initiation or shuddi sanskar, and promised to live up to the ten governing principles of the sect. The center of the movement was then Chaguanas, and it spread into Naparima through Suruj Balli in San Fernando, and Baldeo Persad (Shastri) in Débé. The movement has never been strong in San Fernando, largely because of the competitive position of the Presbyterians.

The attraction of Arya Samaj is its rationalistic outlook; it emphasizes equality not caste—it is not priest-ridden; and women have a prominent place in the movement. It appeals to Hindus growing up in a westernized community. So why has Arya Samaj not caught on? The archaic Sanskrit of the vedas necessitates some training, and Sanathanist priests are the obvious candidates. Other weaknesses have been Arya Samaj's lack of organization and shortage of funds.

The influence of Arya Samaj has been greatest through reforms it has brought to traditional Hinduism: day weddings; the marriage ceremony shorter, simpler, and easier to understand; there is no longer a dowry, only a voluntary gift during the marriage ceremony; visitors to the wedding are now able to eat together at the common table. It is significant that the first two Hindu primary schools in Trinidad were set up by the Arya Samaj in Chaguanas and Curepe, and Arya Samaj was the first non-Christian organization to receive a government grant for building a school. The Ecclesiastical Grant now assumes there are 6000 Arya Samajis in Trinidad, but Mr. Lal puts the number of adherents at between 6000 and 10,000.

Mr. Lal is an Arya Samaji pundit, but prefers not to practice as a priest. He has also been a licensed marriage officer since 1955. Licences are issued by the minister of home affairs. Merit, not caste, is the basis for selection.

The national governing body of Arya Samaj is the Arya Pratinidhi Sabha. The Sabha is made up of branches; it controls all the property and has an educational board, of which Mr. Lal is the secretary. Mr. Lal has been past president and secretary of the Sabha (1946–60), but the Sabha is now split. The disagreement is about whether all members of the branches are members of the central body, since this has implications for voting. Originally, every branch had two on the central committee, and voting was confined to that college. The issue has been taken to court, and will not be decided by the body itself. This fine point of law, in Mr. Lal's opinion, is being exploited for political purposes, as each national political party tries to control the non-Sanathanist vote (be it Arya Samaji, Muslim, or Cane Farmer).

Ramnarine Seereram Maharaj (brother of the contractor and cane farmer) is pitted against Norman Girwar. Also involved with him are R. Dhaniram, a laborer at Texaco, Forest Reserve, and Moosai-Maharaj, a worker in hydraulics and brother of Stephen Maharaj, the DLP politician and leader of the opposition. The father of the Moosai-Maharajs was a pundit who became a catechist in Susamachar Church in San Fernando. Dhaniram and Shiva Persad (a lawyer) have been competing for the headship of the Sabha, and Mr. Lal was a compromise president in 1960.

The split in the Arya Samaj and the Cane Farmers Association is identical. Norman Girwar supports the G. B. Lal faction; Seereram Maharaj is behind Shiva Persad, and crucial to the split in the Cane Farmers also. Lawyers in Port of Spain say they cannot decide where one controversy ends and the other begins. Girwar is pro-PNM; and Seereram Maharaj is pro-DLP.

G. B. Lal is close friends with Norman Girwar and his wife, and with Frank Misir, and he knows Rajkumari Maharaj and Walter Annamunthodo.

There are two main annual functions celebrated by the Arya Samaj: in February/March, Rishi Bodh Utsov (or Siw Ratri), when Swami Dyanand, the founder of Arya Samaj, rejected idols; and in October/November, Nirvana Utsov (on Diwali night). Last year, Nirvana Utsov was celebrated at Débé and Avocat near Fyzabad.

Arya Samaj recognizes 16 sanskar from birth to death, and Mr. Lal mentioned some of the more important ones:

- Garbardhan Sanskar (conception)
- Namkeran Sanskar (naming on birth)
- Anaprashan Sanskar (first solid food)
- Viva Sanskar *(marriage)*
- Antyeshti Sanskar (death)

All Arya Samajis can take the *janeo*. Swami Dyanand's book, *Sanskar Viddhi,* describes how all the sanskar are to be performed. For example, the dead should be cremated, and a *hawan* lit at the cremation (or graveside, if there is a burial). This should be followed by a purification ceremony at the house of the deceased.

Arya Samajis are better read than Sanathanists, and are more familiar with the terms used, and literature referred to, by Mr. Bhattacharya in his Gita class. Mr. Lal remarked, "You cannot go back to India. You should have churches and regular services like Christian people."

"Hinduism will fizzle out, unless they all meet and work out a compromise. Sanathanists will not get the young people. We are not steeped in the religious atmosphere of India. We know what money can do. The next generation will determine what Indians will contribute to the melting pot of Trinidad." Young Indians sing catchy tunes, such as the songs of Hemant Kumar. Indian songs on the radio are listened to by some Creoles. But, in the future, Hindi will be lost. Sanskrit is used only in the rituals of the Arya Samajis and the Sanathanists. Yet, "Indians have something to contribute—if they think about it."

"Indians are thrifty: marriage is followed by having a family and building one's own house." But now Mr. Lal's daughter is in London, and so are his brother and other relatives. "Indians, in general, however, have been too conservative, and have not gone out of their way to meet blacks." Mr. Lal's wife is black and Roman Catholic, and they have more non-Indian friends in San Fernando than Indian. His wife cooks vegetarian food, though Mr. Lal takes whisky occasionally and meat, excluding beef and pork. On the whole, he thinks that intermarriage is not good, since it enhances conflict and encourages a dominant will to win.

Social mixing in San Fernando is carried out through functions, rather than on a house-visiting basis. Presbyterianism is a closed circuit. Hindu social gatherings are religious functions; Hindus have family, not friends. Muslims mix socially more than Hindus.

The Todd Street *mandir* was preceded by a wooden temple that dated back to 1940. The Prince Albert Street temple was on land owned by Mr. Rienzi;[47] at one time, Mr. L. Jaggernauth MBE was secretary of the Temple Committee, and the *pujari* was Ramsamooj (not Hansar).

"Caste is not strong among Hindus, though there is a strong 'up' feeling among Brahmins. It will go into the next generation. They are clannish in comparison with British Guiana, where the wives of some priests were originally Muslim. Here you often see the feeding of the Brahmins first."

Dr. Rampaul, Princess Margaret Hospital

Dr. Rampaul (whose father is the president of the Seunarine Sabha) is a doctor at the Princess Margaret Hospital in San Fernando. He attended the Eckel Village Canadian Mission School, and then went on to Naparima College, where he took his Senior Cambridge. Supported by his father, he left for India in 1956, studied at medical schools in Bombay and Amritsar, and returned to work in Trinidad in early 1964. His sister also has a BA from Bombay in English, Logic and Psychology, and his brother is currently studying medicine in Bombay. These children have been educated at Indian universities, because, being East Indian, their parents were keen and interested in the connection; furthermore, it is cheaper to study in India than in the United Kingdom or United States.

Having been educated at Christian schools, Dr. Rampaul is not particularly interested in Hinduism, and found India rather disappointing. There was so much poverty, and it was such a struggle for existence. City people were not as hospitable as he had expected. Food was a letdown—it was all peppers and spices. The climate was trying, with temperatures in Amritsar rising to just below 45°C. And the multiplicity of languages was confusing. Fortunately, the students were friendly, and there were many from overseas—Germany, Russia, the Middle East, and Africa.

Dr. Rampaul knows some Hindi, and learned Punjabi while he was in India. It now strikes him that Trinidad Indians speak a broken Hindi. Many of the ceremonies performed here are from 200 years ago, and are not found now in modern India. There are no *jhandi* in India, except outside the temples. In India, people go to the temple or the riverside; religion is more domestic here. In India, the groom rides on a horse and not in a car in the *barat*. Night weddings are the norm in India.

Mixing with Indians in India was difficult because of the language problem. Even setting that on one side, it was difficult to identify with them. Most young Trinidadians going to India are disappointed. They will have traveled with the idea that everything in the garden is rosy. The older generation painted a glowing picture; but Trinidad has a western atmosphere. East African and Mauritian Indian students have kept more of the Indian languages, and mix more easily than the Trinidad Indians. But Dr. Rampaul's reaction on returning from India was that, "Trinidad is a little paradise: the buildings have changed, but not much otherwise."

He is not interested in his father's Seunarine ceremonies. He considers himself a Hindu but does not affiliate to a particular sect. In

India, Seunarine Hinduism does not count; there is only Hinduism. When he was in Amritsar, he used to go to the Sikh Temple or to a Brahmin Temple—all could go to both. In his view, caste is still strong in village India, but in the university, there were no caste distinctions, nor was there discrimination against the scheduled castes.

"I feel that I am a West Indian first and an East Indian after, but probably did not before I went away." This feeling is true only for educated Indians. Dr. Rampaul will marry an East Indian for reasons of adjustment back to Trinidad—not that he is against women of other races. His parents would not object. He believes in love at first sight, so courtship is not necessary. One thing is clear: his father would not dream of arranging a marriage for him. He would like to get married according to Hindu rites, but not necessarily Seunarine rites—he would prefer something more modern. Walking around the sacred fire does not appeal to him. It would be best to call a pundit and get him to sort it out.

Since his time in India, Dr. Rampaul has become more West Indian. Most Indians outside the Indian Ministry of External Affairs have never heard of Trinidad. He eats anything, and feels quite at home, if not more so, with Creoles. A year or two back, politics (presumably the 1961 election) had caused some racial antagonism in the hospital, and one senior nurse is definitely anti-East Indian. But most black colleagues invite Dr. Rampaul to their parties, and he has many black friends, both in the hospital and at home. One of his classmates at Naparima College was black. In a class of 36, only 6 were non-Christian. But that was a better situation than when his father was at school, and his path toward the teaching profession was barred because he refused to become Presbyterian.

Creoles like to dress up and spend all their money at the weekend. Indians are thrifty and think of their children and grandchildren, and this is not true of all peoples. Dr. Rampaul's father and two older brothers own a sawmill. His father rented a house in San Fernando for Dr. Rampaul and his four brothers and sisters to live in. They looked after themselves for five years while attending local secondary schools.

Wednesday, August 26

Colin: *Mr. Sooknanan Maharaj, property owner and businessman*

Sooknanan Maharaj, aged 71, was born in Vistabella on the northern outskirts of San Fernando in 1893. His father was an indentured laborer on Reform Estate, now Usine property, though he also did his own gardening. Mr. Maharaj was brought up by his grandmother. He did some roadwork, and at the age of 18 his grandmother married him off. By 1924 he was doing cane planting and gardening at Hermitage, producing 20 tons of cane plus rice; then, in the early 1940s, he became a market vendor and raised corn, cattle, and mules.

Eventually, he bought a 35-acre property from Stollmeyer, but had no money to put in. In 1953 he built a house, and completed the Vistabella *mandir* near his house in 1959–60. It is open to the public, though Mr. Maharaj still maintains it. There is no deed or written authority.

Services take place now and again—especially the Sunday School and sandhya puja, usually under the direction of Moon Pundit from La Romaine. The temple can house up to 100 adults. They celebrate Diwali (festival of lights devoted to Lakshmi), Siw Ratri (celebration of Siva's birth), Jhanaam Astamie (birth of Lord Krishna), but never Phagwa—you have to go to Débé or San Juan (Port of Spain) for that. There are plenty of Indians here, but no strong leaders. On a Sunday in the middle of the month, Brahmachari Hari Ram takes the service, and Mr. Maharaj wants to hand the work of the temple over to him.

Mr. Maharaj designed the temple himself, and used four different builders. It is dedicated to Shiva (God the Destroyer). A *saddhu* looks after it, and offers prayers morning and evening. Moon Pundit is none too interested. "Some say, it's yours, and will not come." A few get married in the temple. There have been four so far, and there is no charge. Hari Ram wants to try and develop a secondary school. But Mr. Maharaj does not want him to go to a denominational body, such as the Maha Sabha Board, for funding.

Marriage is "hard if you have to look around for a day's work." But you cannot stop children from taking a bride of low caste or white American or European. "They have to study their own future"—but he would not like a black. "If they like it, they keep away from me. It could be the governor, but me no like. Blacks are beasts: we can live together, but home business altogether different. We should do business together; but not at home. A low nation (caste) is better than blacks."

The house has religious pictures, and a photo of Mr. Maharaj's son, Lal, leading a winning racehorse into the paddock. Mr. Maharaj is a good friend of Binie Maharaj, who is regarded as family—his cousin is a brother-in-law of Mr. Binie. They do the *rōt* trilogy (Hanuman *puja* on Saturday morning with a red *jhandi*, Satnaryn[48] *katha* on Saturday afternoon with a white *jhandi*, and Suruj Puran[49] on Sunday morning also with a white *jhandi*) once or twice a year; and a Lakshmi *puja* (yellow *jhandi*) at Diwali and before his wife goes on holiday to the United States and Canada to visit family.[50]

Mr. Dialdas, a family friend, is planning to bring his father to San Fernando from India. Mrs. Dialdas is bye-family, a cousin of Mr. Maharaj's wife and born in San Fernando (whereas Mr. Dialdas is from Bombay).There is also a photo of Mr. Maharaj's daughter in a sari seeing off friends at Piarco Airport.

Thursday, August 27

Colin: *Rotary Club Luncheon*

Frank Misir invited me to a Rotary Club Luncheon (an all-male affair) at Naparima Club. The president, Peter Hadden, was in the United Kingdom, so Frank took the chair as acting president. Out of the 23 present (excluding me), 12 were white, 7 Indian, 3 black (of whom 2 were very light) and 1 racially mixed. Two of the Indians were guests, Mohammed Shah[51] (barrister), and Mr. Lalla (barrister). This was a fine example of elite multiracialism, with everyone knowing everyone else, and all at ease.

Alloy Lequay on business clubs and politics

After lunch, I went on to have a second conversation with Alloy Lequay. It turns out that his sister is the wife of Chattoor, whom I had just met at the Rotary Lunch. Talk turned to other San Fernando clubs, the Junior Chamber of Commerce, and the Lions Club, in both of which Lequay is involved.

The Junior Chamber (JC) has just been started in San Fernando in March 1964, and focuses on community service through leadership and training. The president is David Hahn of Alstons, and Lequay is the second vice president. Previously there was just the Trinidad JC, really focusing on Port of Spain. Most of the members are junior executives, and they have to be aged under 40 The chamber, with a membership set at 50, meets at Naparima Club. It provides training courses for its members in business administration and parliamentary procedure.

The Lions also give service to the community, and organize projects to raise funds for charity. The San Fernando Lions Club, which is only two months old, has 20 members (the minimum is 17), and they are mostly older than the JCs. Lequay gave me the names of 15 members: 7 were Creole, 6 East Indians (including the president), 1 Chinese, and 1 Chinese Indian.

In the second part of the conversation, we turned to Lequay's involvement in DLP politics. Once more he reminded me that he is more interested in sport than in politics. But he immediately added that the DLP has a poor record because of its lack of organization, and it is too early to forecast the next election. There is a great problem in establishing a stable organization and creating trust in the ordinary man.

Votes come mostly because of community work, not on a race basis. Saied Mohammed got the hardcore PNM vote and Muslim support in San Fernando West in 1961. As far as the elections go, it is just a case of saying we shall do better. There is the problem of creating small party groups and organizing discussion at the local level.

In 1960, Lequay was asked by Ashford Sinanan (then acting political leader) to assist in organizing the DLP. He had never been in politics before. He could have helped the PNM in 1956, but did not; however, he thought that help was needed in the DLP, and so he responded to Sinanan's request. The first time he was on a political platform was in 1961. He thought that the party could not win more than 10 or 12 seats,[52] though he estimated that he himself had a fifty-fifty chance in San Fernando West (he lost to Saied Mohammed). Lequay is now the DLP party secretary; it is difficult to do that and be up for election.

The older political leadership will hang on because they like power, though they do not like Capildeo's policy of democratic socialism, preferring the anachronisms of Bhadase's days. The rank and file will not accept Bhadase, and Bhadase will not come in at any level other than the top. Lequay does not think the 1961 elections were rigged by the use of voting machines, though they did not function properly.

San Fernando is much more racially tolerant than the country as a whole. At the last election one could hold political meetings, and Lequay did so, till the last minute, as did Muradali (they were the two DLP contestants in San Fernando). The racial situation has been better since independence (in 1962). Trinidad is no British Guiana—but racial tolerance in Trinidad depends on the leaders, and they are respected by both communities. Yet Bhadase could split the country, possibly by forming a party within a trade union, which could lead to another British Guiana.

Friday, August 28

Colin: *Robert Montano, businessman*

Robert Montano was interviewed by me during the afternoon in his office at Imperial Stores. He declared himself fascinated by the multiracial set up of Trinidad. Race and politics involve fishing in troubled waters; otherwise, the groups live side by side in considerable harmony, though the lower social levels in San Fernando should be excluded from that favorable comment.

Robert Montano was born and bred in San Fernando, and educated at St Benedict's, now Presentation College. He joined the Canadian forces to escape the world of business and his tyrannical father, and spent World War II flying Lincoln bombers before rejoining the firm in 1945. His first impression, on his return to Trinidad after the war, was that everything was small and dirty; he had held idealized memories.

Robert had a number of East Indian friends at school, and almost half the staff in his business—Imperial Stores—are Indian. Whether working or playing, some of his closest associates are Indian. "I am fortunate in being neither one thing nor the other." He is white-Chinese. His wife is English, and he has acquired most of his business friends through Rotary.

The French Creoles are not as powerful as they were a generation ago, when the Maingots, Rostants, and De Verteuils were socially dominant. There are a few top blacks, especially in Port of Spain, where the civil service has been a springboard for upward mobility. Indians are moving from their association with the cane to top positions, "but we regard San Fernando as harder-working than the capital." Port of Spain suffers from a fête mentality, and blacks, from an ancient ill-treatment at many levels. Blacks are not at all hardworking, and many people say that "monkey keeps bringing up slavery." The fact is that Williams has not exorcised this ghost.

Blacks have found security in what Williams has said, but have misunderstood it—hence their truculence. The fact that the black has no respect for authority can be traced back to Williams' speech, "Massa Day Done." East Indians are brooding, and then hatred erupts. Black crime is quite different. East Indians in the popular imagination are fond of litigation; but Indian lawyers take the line, "if you can't beat 'em, join 'em." For poor Indians, a professional training is *the* status symbol, even though parents might not understand what education for their children will really mean.

It is a commonly held belief—probably held by Indians too—that Indians are weaker physical specimens than blacks. Some say that the East Indian is untrustworthy; never thrust him into important positions, because of the risk of *bobol*;[53] but the Indian is shrewder than the black. Pride in being black goes back only to Williams (1956), whereas the East Indian has always been proud of his race. "I have the highest possible regard for the East Indian. There is a fear that the Indian will outbreed the black and assume political dominance. Rural East Indians, especially, will vote race."

A few years ago, Indian leaders used birth-control slogans in reverse (exhorting Indians to reproduce). They felt justified for feeling that the possible incorporation of Grenada into Trinidad and Tobago was a trick to keep up the black voting strength. Robert does not think Williams himself is "in the least bit racist," but he certainly knows that his strength lies with the blacks. Few blacks would vote DLP: it would be "treason." Williams "has created black unity and the idea of the black intellectual." Blacks would vote Liberal rather than DLP.

Does Trinidad have the makings of another British Guiana? The elements are there; it depends on the evolving pattern of leadership. If a rabid Indian leader took control of the DLP, or the same development occurred in the PNM, then Trinidad could become another British Guiana. What can the business leaders do? "I think we should speak out against excesses on either side."

Robert is a close friend of Jack Kelshall.[54] In 1953 a peace rally was held at the Town Hall (organized by the Communist Party). Gerard (Robert's brother) was then on the town council, and Robert was invited to participate because of his views on social justice. In the business community, there was real deep-seated fear. All the participants in the rally were asked to stand and sing the Internationale. Robert said to the assembly, "I know the value of peace because I have seen war." They never held another public meeting. The business community is very conservative, "but I am not so pink."

Eight years ago, Robert was a moving figure in the establishment of the Southern Chamber of Commerce in San Fernando, and is currently the president. He commented on the great difference in racial composition of Port of Spain and San Fernando and on the lack of integration of oil and sugar in San Fernando, although it is the center for both. There is conflict over so many things, but the East Indians will make San Fernando progressive in two generations. The Chamber now has 150 members reflecting the sugar, oil, cement, and chemical industries and the merchants of San Fernando, who make up a large percentage of the membership. The Southern Chamber of Commerce is pushing for an industrial estate and deepwater frontage.

In Robert's opinion, government policy over birth control should argue that "large families are not in the interest of the nation." Before Williams went into politics, he and Robert were, in 1954, in a study group of 15 discussing population policy. But the Catholic Church brought Williams to heel over the issue at the 1956 election—a great mistake according to Robert.[55] "The Catholic Church can bring forward outmoded arguments in this little backwater," and then there are the views of the Muslims and Hindus (Robert admits that he knows very little about Hinduism). Catholic priests used to put forward the idea that a small family was positively bad.

The French Creoles got together with the expatriates from the estates to form Naparima Club about 40 years ago. It was exclusive then. But it has had a lot of competition from the clubs of some of the big companies. Now, "mixing is something we in San Fernando are very proud of." Naparima was once the stronghold of the whites, but over the last three years or so there has been a big change; now people are scrambling to get in, and this is having an important influence on the community. Robert would not object to his children having black or Indian spouses, provided they were "charming people."

East Indians are more acceptable to whites than blacks are. For their part, Indians have a strong preference for whites, because of the whites' social position and their birth links to planter ancestors.

Dr. Abdool, hospital doctor and on ASJA Executive

Later the same day I had an interview with Dr. Abdool. Dr. Abdool was born into a Muslim family in Fyzabad, but in 1952 the entire family moved to San Fernando. Dr. Abdool attended Presbyterian Primary School and then Naparima College, where he was awarded his Senior Cambridge Certificate. However, his father, a taxi driver who had got into bad company and was going down socially, could not afford to support him at university.

Between 1950 and 1954, Dr. Abdool taught with the Anglican Board at Forest Reserve, east of Siparia, where he was put under some pressure to convert, and later at Point Fortin and Marabella. But in 1954 he left for Canada and started to study at McGill University, working over the summer in a goldmine and doing babysitting. In 1955, faced with the problem of getting into medical school, he went to the University College of the West Indies in Jamaica, took Physics, Chemistry, and Biology in 1955, and in 1956 started medicine. In

1955 he became the first president of the UCWI Islamic Movement, which attracted ten or so students on a regular basis. I met him in Kingston while I was carrying out fieldwork for my doctorate in 1961.

Dr. Abdool feels himself to be more Islamic than Indian. He likes Indian music and food, but dislikes Indian shows. Dr. Abdool is the son-in-law of Mr. Shafik Rahaman, the brother of Sheik Rahaman. He is on the ASJA Executive Committee, the Executive Secondary School Board (with Wahid Ali,[56] whom I also met in Kingston in 1961), the Executive Primary School Board, the Council of Muslim Youth Organizations' Executive, and the Executive of the Young Muslim Cultural League. San Fernando is the backbone of ASJA; its *jamaat* (congregation) is one of the few to be well administered.

Dr. Abdool finds it impossible to keep up the five daily prayers, though he does carry them out in the morning, at sunset, and at night. He fasts for Ramadan, but cannot leave the Heart Clinic for Friday prayers in the mosque, even though it is obligatory.

The Muslim family environment is conducive to parental control of marriage partners, though the couple might be able to go out after their engagement. In Dr. Abdool's view, Muslims are becoming more modernized. Muslims have a long way to go before they become anything like as modern as the Hindus.

Gillian: *Will Hercules on San Fernando society*

Will Hercules, our black barrister friend, came round in the evening and told us what he knows about San Fernando society. According to Will, French Creoles and the English were, historically, the cream of San Fernando. There are still about ten families in the elite, split equally between French and English.

"French Creoles think themselves better than the Montanos; the Montanos think themselves better than the Dieffenthalers; the Dieffenthalers, because of their light pigmentation, think they are better than the Scotts." However, the Scotts really belong to the group of blacks who entered the elite in the period 1919–39. They were one of eight families to do so, and were matched by a similar number of prominent East Indian families who joined the elite at that time.

The lists he gave us show that 11 "old" Creole and East Indian families were still prominent by 1964, but their elite status was being contested by new arrivals, both East Indian (50 families) and Creole (35 families). Creole newcomers were involved in politics, the civil service, medicine, and the law, while East Indians were particularly prominent in business, medicine, and the law.[57]

Will claims: "I get on splendidly with East Indians; an Indian will come to me as a client, because they do not trust their own people." If he comes up against deceit in court, Will reminds the Indians that "they are coolies." Will believes that the Indians are working hard to take over, and are moving into top professional and other positions. Blacks are conditioned by harsh economic factors that Will likens to slavery. In the same breath, he dismisses blacks as "interested in filling their bellies, drinking rum, wearing clothes and playing mas." But he is convinced there will be no breakdown as in British Guiana, since the Trinidad black is too dependent on Indian-created jobs.

Will's mother, a member of the Dottin family, was a long-haired Carib from Martinique. His father, who taught at Naparima College, was the first person in Trinidad to take a London University External BA degree.

Mr. Gopie and Indian music

Later that evening Colin and I listened to the Government Radio Program, and heard an interview between Amjat Farzan Ali (a public relations officer) and Mr. Gopie on the theme of the National Council of Oriental Music and Drama (NCOMD). Gopie spoke about the formation of the organization to promote oriental songs, to engage in moral upliftment, to help the disabled, and foster universal love and brotherhood—all Gandhi-inspired.

A British Guianese team is here for a singing contest at Skinner Park. It consists of six people and is getting full government support, unlike the Trinidad team. Mr. Gopie claims he has been "in the field of Indian music and singing for a very long time." There is also going to be a competition for the best song composed on the theme of "independence." Mr. Gopie anticipates large crowds in San Fernando, because the event will provide "very popular entertainment for the Indian community." Last year this was a local affair,—this year international.

Interviewer Ali tells us this will be the "greatest Indian competition ever in Trinidad." Last year over 20000 people participated; this year's event will take place in Skinner Park, San Fernando at noon on Sunday, August 30. Both the interviewer and interviewee gave their audience the impression that NCOMD is well established, and they certainly did not mention that this is its first activity. Gopie introduced himself, the president, as "the humble speaker."

Saturday, August 29

Colin: *Robert Gunness in Mayaro*

We drove back to Mayaro to interview Robert Gunness. He expressed surprise that caste was of any significance among Hindus in San Fernando. Although his maternal grandparents were Muslim, he hints that he is Dhobi (washerman caste). In his view, the leading Hindus in San Fernando are nonentities. Rama Maharaj had a rum shop, now a burnt lot, opposite the market on the corner of Mucurapo Street and Keate Street. Rama is the brother-in-law of Moon Pundit. They are both now contractors. Ramnarine Rampaal Maharaj and Lal Bahadoorsingh were moneylenders, charging 2 percent per month compounded. Union Hall real estate was developed by them. They paid BWI$64,000 (£12,800) for ten acres, and Robert bought one-and-a-half acres from them for BWI$5,400 (£1,080), but, despite their obligation to service the plot, it had no water, electricity, or roads. Robert calls Ramnarine Maharaj "Ramnarine Chamar."

Gerard Montano[58] joined the San Fernando Borough Council in 1946. Roy Joseph was looking for a safe vote to ensure his election as mayor. In the process, Norman Girwar was overlooked, hence his support for the PNM today. Pitty Mahabir, Bram's uncle, has been supporting the PNM to ensure the awarding of road contacts. His green American car was loaned to Saied Mohammed for the 1961 election campaign. Robert thinks that many Indians are PNM. Gerard Montano was launched by Roy Joseph. Both Gerard Montano and Winston Mahabir have Canadian wives. Joan Mahabir was won over by Williams, and the rest followed. It created the impression that professional Indians (especially the Presbyterians) were pro-PNM. Mahabir buried Roy Joseph politically in 1956.

Gillian

It's interesting that Mr. Bhattacharya has been financed and sponsored by Gopaul, Gopie, and Jang Bahadoorsingh. All are PNM! It is their aim to get hold of Mr. Bhattacharya to legitimize their control over the Indian community. Gopie certainly looks to Mr. Bhattacharya to endorse all he does. The PNM seem to have a hold on Hemant Kumar; all his performances are said to be for the party.

Robert told us that the temple at Tableland had been built by Chamars, and that pig sacrifice was held there. The Nanans did something similar at Débé. Robert made the association between Kali Mai and Siparu Mai, in the same way as Dr. Abidh did earlier.

Monday, August 31 (Trinidad and Tobago Independence Day, second anniversary)

Gillian: *Frank Misir, barrister*

Frank Ramnarine Misir[59] was born in 1923, grew up in Princes Town, and was named after the local luminary, Dr. Frank Mahabir. Educated at Naparima College, he took his first bar exams in Trinidad and his finals in the Inns of Court, London. Misir is a caste name, adopted in London because of the problem of distinguishing between various Narines.

Frank's paternal great-grandfather, Bissessar, was from Jaunpur, on the north bank of the Ganges. He came to Trinidad as a professional cook in 1865, and was not indentured; on his immigration papers it said he could travel anywhere he wished on the island. Ramnarine was the name of Frank's grandfather. Old Bissessar did not carry his caste name of Misir. All the letters going back to one of Bissessar's sons still living in India, around 1888, were addressed to Misir (Bissessar had concealed his Brahminical status when on board ship).

Frank's father, and most of his family, became Presbyterian. Frank, however, was formerly a member of Christian Endeavour at Princes Town, a smaller and even more conservative religious society. But he never became a baptized Christian because, despite having always identified with them. The local minister one day tackled him and said, "Why don't you really join us?" From then on, he had less and less to do with them, because he felt rejected and insulted, and so he became an Arya Samajist. Two others who were deeply involved in Arya Samaj at that time were Makhan Dubé (Dubé is a Brahmin subcaste) and Makhan's father. Makhan was taught how to become a pundit, and Frank remained a staunch Arya Samajist until he went to England. Since his return, Frank has found them too dogmatic.

Back in Trinidad from the United Kingdom, Frank married in 1949 (a Samaroo, from the family of sweet drinks manufacturers). Their wedding was according to Sanathanist rites, and was performed by Jankieprasad Sharma. But it was not a normal Hindu ceremony. They were married very simply at a table, "like an Arya Samaj wedding."

In the past many people claimed falsely to be of the Brahmin caste. In former times they were flogged or tried by *panchayat*. Nowadays you can't tell whether people are telling the truth or not. In the late 1880s, *panchayat* decisions were regarded as legal justice in Trinidad. *Panchayats* are still held, but mostly for religious matters; for example, recently in Tableland, when there was the matter of selling cows to a butcher.

Swearing on a *lotah* in a court of law is symbolic of swearing on Ganges water. Now, for most Indians, the original significance has gone, and the *lotah* has no meaning. In today's paper, there was a suggestion that Hindus should swear on a copy of the Gita. When he took his oath as a barrister, Frank made an affirmation. He is deeply concerned about the reputation of Indian lawyers in Trinidad. The best thing they can do is be honest. Too many accept bribes, and are regarded with suspicion by the rest of the community because of this.

We talked about the leading Indian families. David Mahabir from Princes Town was of the fisherman caste (either Mallah or Kewat), and he married a Chamarin. They had two sons, Jules (the first Indian barrister) and Frank (one of the first Indian doctors). They were all Presbyterian, including David. Winston Mahabir, the former PNM politician, is the son of Jules. Other distinguished Indians were the Kangaloos, Namsoos, Fitzpatricks, and Roodals, all of Madrasi origin. Color is more important among Indians than among blacks, so Frank claimed, and a high value is placed on having a white husband or wife. Frank told us about the types of Indian who are widely regarded as rogues: "beware the black Brahmin and the fair Chamar."

Many Telis (oil pressers, traders, and cultivators) became Muslim. The skulls of Telis were used for magical practices, and Frank named several people who had made the same name/religion change, presumably to conceal their caste in ways that simply becoming Christian would not have done.

V.S. Naipaul's *Mystic Masseur* is not straight Ajodha Singh,[60] but a mixture of Ajodha Singh, Doon Pundit, and various others. There seems to be almost universal admiration for Ajodha Singh (even from the likes of Robert Gunness). When he was minister of works, Ajodha Singh had part of his office set aside for patients.

Frank clearly feels that there is no place for him in Trinidad—he is hankering after the political freedom of England, and is thinking of getting free of Trinidad politics and moving to the Seychelles. He feels there is no room for any real personal commitment in Trinidad, and is disillusioned with both the PNM and Capildeo. Formerly, he was a supporter of the PNM, pro-Williams, pro-independence, but he admits that he was unable to foresee the drawbacks.

Frank thinks that there is no visible racial prejudice shown by blacks toward Indians, but there is palpable prejudice. In our view, his situation is very like that of Will Hercules. Both are disillusioned with Williams, and both are caught up in a racial situation. Both see the Liberals as the answer, though they are seriously hampered by lack of support. A problem for both Will and Frank is that of knowing where the allegiance of their colleagues lies. No matter what each does and says, Will is always regarded as a black, and therefore PNM, and Frank, as Indian, therefore DLP.

Frank believes that wherever Indians go, they make trouble, for example, in Mauritius. They are "objectionable," that is, they bring problems of diet and suchlike. Krishna Menon, the former foreign minister of India, would have nothing to do with descendants of "slave" Indians overseas, even though those very descendants were agitating in London for India's independence. East Indians who came back here after Indian independence in 1947 lost interest in politics completely.

But blacks, for example, David Pitt,[61] who were never really involved in London, came here and used the issue as a springboard to go into politics in the West Indian National Party, which also involved Patrick Solomon[62] (now minister of home affairs) and Garvin Scott (now High Court Judge).

Frank's wife follows no dietary restriction. At the Rotary lunch I attended, Frank was served chicken instead of the beefsteak that the rest of us had on high table. Frank sometimes eats pork, however. One day, when Frank, as a young man, was working at the U.S. base at Chaguaramas and had no control over his diet, he found bones among the string beans, and questioned the cook about it. She replied, "Who you fooling man? All you coolies does sleep with nigger woman, and you still bother about your food!"

On one occasion, Tajmool Hosein and Mohammed Shah, sitting on either side of Frank at a Bar dinner, were served pâté de foie gras. Shah ate his and the others held back. When Shah had finished, Frank leaned over and asked if he knew that he was eating pork. Unmoved, Shah replied, "Hosein see?" Frank said, "Yes." The issue was not whether he had eaten pork (it was in fact goose liver), but whether he had been spotted! Many Hindus and Muslims, who eat and drink everything in private, make a point of not doing so in public. Mohammed Shah was formerly associated with ASJA, but not now. (Mohammed Shah was Lessey's office boy; Chattoor was Hobson's office boy.)

According to Frank, the clannishness of Hindus in the 1940s and early 1950s was the breeding ground for the involvement of race in politics. Ranjit Kumar was supposed to be canvassing the Indian vote for the Trade Unionist, McDonald Moses,[63] an associate of Rienzi. But Ranjit Kumar[64] was told by the East Indians in County Victoria[65] that if he went up on an independent ticket, they would vote for him as an Indian. Kumar rode around on horseback dressed up as a Hindu deity, and this was the start of trouble.

Colin

Frank told us that in the late 1930s and early 1940s, Rienzi had the solid backing of both Indian and black workers. He was president of the Sugar Workers Union, the Oilfield Workers Union and the Trades Union Congress, all at the same time. The visit of Dr. D. P. Pandia from India in the early 1940s brought Rienzi back into the "Indian fold." Some blacks were already trying to outmaneuver him; eventually he accepted a government post in the law and became sidelined. Nevertheless, many blacks claimed that this was an example of Indian deception; and even Bram told us that he regarded Rienzi as selling out to the colonial government.

In 1950, the Butler Party[66] got six seats in the Legislative Council, and should have been asked to form the government. But the Butler Party was overlooked by the governor in the formation of the Executive Council, and the two Sinanans (Ashford and Mitra[67]) deserted the party. Blacks decided never again to trust East Indian candidates.

The West Indian National Party was waiting for dynamic leadership, which Eric Williams supplied—to the PNM.

In 1950, Bhadase Maraj was elected to the Legislative Council. This marks the beginning of the political strength of the Maha Sabha. The only Indians put up by the Maha Sabha were Hindu. This led to the alienation of Muslim voters, and prepared the ground for 1956, when Williams used this to his own advantage. The Maha Sabha did put up "an apology of a Muslim"—a rum-shop owner. Contrast the success of Kamaludin Mohammed in the Williams government, and of Olie Mohammed, mayor of San Fernando, 1956–57. The Sinanans joined Roy Joseph when they left Butler. Frank comments that in his youth he saw the races split by political activity in the Princes Town area.

Among upper bracket East Indians, no one can be sure of the political allegiance of the next man. For example, Norman Girwar is a staunch PNM supporter, because he was overlooked by Roy Joseph in favor of Gerard Montano for a seat on the San Fernando Borough Council in 1946. Roy Joseph went up on an Independent ticket (with Maha Sabha backing) in 1956, but was beaten by Winston Mahabir.

Roy Joseph had previously been sure of the Muslim vote because his wife was Muslim, and he had many friends and contacts through her. (Dr. Abdool said that when Roy Joseph was minister of education, the Muslims got school boards going, and he was very helpful to Muslims—including Abdul and Shafik Rahaman—in the formation of primary schools.)

Frank Misir was the first person elected to office in the San Fernando Rotary Club, which was originally started a few years ago by expatriate whites. Frank does not mix with the Rotary group, other than with his legal associates. His closest friends are "a mixed bunch" from Princes Town.

Wednesday, September 2

Colin: *Sheik Fuzloo Rahaman, President ASJA, San Fernando*

Sheik Fuzloo Rahaman was born 51 years ago in Mucurapo Street. His father, who was born in 1876, was the imam of the Jama Masjid Mosque (built in 1913), located next door. It is claimed that he was the first Trinidad-born Indian to go to India—in 1888. He studied for ten years and became a *hafiz*. Fuzloo was taught by his father until he was 12, and married Mr. Jaleel's daughter, Zobida, when he was 23. His brother, Shafik Rahaman, is now 46. After more than a decade of refusing office, Fuzloo is now the president of ASJA (incorporated in 1935) in San Fernando. "The hub of Islamic affairs since the time of my father has been San Fernando." His mother's brother is Mr. Hosein, the current imam.

Fuzloo recites Arabic but does not understand it. However, he is able to read and, to a lesser degree, speak Urdu. Mouloud Sharif is normally said in Urdu and then translated into English. One or two children read Arabic and Urdu, but they would not really understand. Islam is more vibrant than 20 years ago; but 20 years ago, there were fewer distractions. ASJA has seven primary schools and two secondary.

Mr. L. Jaggernauth, businessman

Like Fuzloo Rahaman, Mr. L. Jaggernauth[68] is still living in the house of his birth—9 Prince of Wales Street, San Fernando, close to the Mucurapo Street Market. He told me that the last shipload of indentured laborers arrived aboard the *Ganges* on April 22, 1917; the last return ship for the so-called free passage on October 22, 1936 was also the *Ganges*.

Mr. Jaggernauth was educated at Grant School (the CMI school attached to Susamachar Church)—named after Rev. Morton's assistant (Rev. Grant), and for a short while at Naparima College. His first job was at La Fortune, as a sugar chemist, but he soon moved on to become a mercantile clerk for Rahamut in the High Street, the largest and best store in San Fernando in the 1920s and 1930s. In 1917, High Street had been almost exclusively East Indian. Mr. Jaggernauth went into business as a sports goods retailer in 1947, and he is now one of the oldest clerks/directors on High Street. In 1949, he was made an OBE, and in 1958 visited India and Europe.

His father had arrived from Uttar Pradesh—near Delhi—in 1892, and was indentured at Union Hall on the southern limits of San Fernando. He had wanted to return to India in 1921. Mr. Jaggernauth's mother was given a piece of land, really just a tract on Prince of Wales Street, which he still owns. Mr. Jaggernauth referred to the Prince Albert Street *mandir*, where they performed *kathas* and Suruj Puran. At the end of the nineteenth century, Mr. Jaggernauth's grandfather, Mohun, took an interest in the *mandir*, as did his father at a later date.

Mr. Jaggernauth is not associated with the Seva Sangh, of which Bisram Gopie is the chairman, or the Todd Street *mandir*: as he said, "some associations don't function." He was responsible for getting the land from Usine in the 1950s. Ramsamooj squats in the old Prince Albert Street *mandir*. H. V. Gopaul was behind the Todd Street *mandir*, but it is not registered with the Sanathan Dharma Maha Sabha. Prince Albert Street functioned until 1950; long ago, there was a Hindu school, but it became defunct in the 1950s. The president of the *mandir* was L. Jaggernauth; secretary, Sharma Maharaj (does pundit work). Others associated with it were Ganga Bissoon, Ramnarine, Bissoon Pundit, Bisram Gopie, and Binie Maharaj.

A Hindi speaker, Mr. Jaggernauth was married by Hindu rites, as were the majority of his children. For religious ceremonies he calls upon Ramnarine or Bissoon Pundit; these are the most important.

According to Mr. Jaggernauth, the 27 principal Indian shops in San Fernando in 1964 were owned as follows: 4 by Bombay businessmen; 3 by Muslims; and 20 by Hindus (5 Brahmins, 4 Chattri, 11 other castes).

Gillian: *Indra Maharaj on Drupati, marriage and her mother*

Indra Maharaj, Mahadeo Pundit's youngest daughter, came to tell us that her eldest sister, Drupati, is faced with the possibility of marriage. Drupati is reconciled to the fact that her parents will look for a husband, but none of the sisters wants a pundit's son or a future pundit. Drupati wants a "social" husband—probably like Uncle Capildeo.

Indra and her sisters all want to marry a Brahmin. She told us the story of a female cousin married to a non-Brahmin (an Ahir). The children are not Brahmin, and are to a certain extent ostracized; for example, five Brahmins are needed for the ritual raising of the *jhandi* flagpole after a Brahmin *puja*, and these Ahir children cannot take part in the ceremony.

Colin

Drupati told me that she would rather go to England now, and be unhappily married later, than be happily married now, and regret not having gone abroad or done something special with her life.

Indra would like to go to India and the United States—to see where the film stars come from. She gets Indian film magazines. Indra regards 21 as a good age for girls to marry; earlier than that is too young. She can hardly believe that our mothers were about 30 when they married. The impression she gives is that, if you manage to stay single until you are 21, you really are emancipated!

Kinship terms, such as *aji* (father's mother) and *dadi* (father's elder brother's wife) are used within their family. Mahadeo frequently calls his wife "*doolahin.*" Indra and her sisters call Capildeo's wife "*tante.*"[69] Indra finds Hindu ways too strict, especially the attitude toward daughters. In a Hindu family, brothers tend to play the role of respectable escort. Indra is at St. John's College, San Fernando, but she does not mix with the boys, or with the blacks.

When Indra met her father's first wife at a Bhagwat in Arouca, she was horrified that she might have had so dark a mother. Mahadeo had married his first wife on the death of his father. She refused to live with him and help to support his widowed mother and young brothers and sisters, and he was too proud to go and fetch her. So, they broke up, and he married his present wife, Ramdoolarie, when she was only 13.

On another occasion, Ramdoolarie told me that she remembers Mahadeo coming to the house when she was playing in the yard. She had no idea that this older man wanted to marry her. She was completely taken aback with fear and incomprehension when she was told that he was to be her husband. As his wife, she became both daughter-in-law with all the duties that entailed and playmate for her husband's brothers. Indra reckons that her mother must have made or started to make 18 babies. Indra weeps when her mother, who is (we think) in her thirties, speaks of the trials of her early married life. Years of childbearing have taken a physical toll.

Thursday, September 3

Gillian: *Robert Gunness and Capildeo Maharaj*

We became involved in a chance conversation with Robert Gunness and Capildeo Maharaj in Robert's surgery. Capildeo does not think that caste is of such great importance, but Brahminism is. Brahminism is strong in San Fernando.

Capildeo claims that young Hindus are chronically embarrassed by the poor Hindu leadership at all levels. Capildeo no longer attends the pundits' *parishad* (council), and his brother, Pundit Mahadeo only goes when it is convenient to him.

Capildeo speaks of himself as the "*kujat,*" that is, the outcaste. Although he married a Brahmin, his wife came from a poor and noninfluential family, and on these two grounds was opposed by his brothers. "I have a wife I am proud to go out with." Capildeo thinks that few men who have had a secondary school education will be content to allow their parents to choose a bride for them. With girls it is a little different still. Robert thinks that *douglas* despise their Indian mothers for having gone to bed with a black.

When Robert Gunness wrote to his mother, announcing his intention to marry Jean, (his English girlfriend), she wrote back and told him that, if he persisted with his plans, he was never to come back to Trinidad. He replied, "I am definitely going to marry Jean. You were the only reason why I wanted to return to Trinidad, now I feel free to do otherwise." She retracted at once, and asked him to come back.

On their arrival in Port of Spain as a newly married couple, Robert's mother took four, specially made, heavy gold bangles down to the ship with her. She presented them to Jean, and reminded Robert that he had married the only child of an English couple and had brought her thousands of miles to Trinidad. If ever he ran around with any other woman, there would be big trouble. Nowadays mother-in-law and daughter-in-law are inseparable. Jean never makes a decision about herself or the children without consulting Ma. Ma relies heavily on Jean for financial support.

Colin: *Mrs. Ivy Gunness (Ma)*

Later in the day we had an interview with Mrs. Ivy Gunness, Robert's mother, known to all as Ma. She was born in 1908 in Williamsville, opposite the railway station. Both her grandfathers came from India. When her father died, her eldest brother, Claudius Niamath, head of Rock River CMI School, who owned land on the Moruga Road, beyond Penal, took the family to live with him. "Indian people make a match for the girls," and in 1924, at the age of 16, he arranged her marriage; she had no choice.

Ma's husband had been brought from India at the age of five by his mother, an indentured laborer. Her husband had been lost in some violent dispute, and she and the children "were running away in the jungle, I suppose." She brought five children, of whom three were known to Ma. They were sent, in 1883, to Petit Morne Estate (on the site of Usine St Madeleine), where the Rostant family were managers. Ma's future mother-in-law and children were put in servants' quarters, not in the barracks, and were never worked in the fields.

The Rostants made sure that Ma's husband, as a little boy, had books and went to school. Mrs. Rostant separated from her husband and she became mentally ill. While the other brothers followed the Rostants and became Catholic, Ma's future husband became a Presbyterian at the age of 10 or 12. As a pupil teacher, he went to Naparima Training College (where he was taught by Dr. Coffin and Dr. Scrimgeour), and eventually became head teacher at Brothers CMI School and at Débé, where he served for 35 years.

Ma's father was called Niamath Ali, and he and his wife were married as Muslims. Both had relatively well-to-do families, but Ma's mother did not like her in-laws (with whom she lived) and she used to run away from time to time. Rev. McRea advised the parents to register their marriage, and they eventually became Christian, without ever knowing much about Christianity. Ma's father was a teacher, cane farmer, and cocoa proprietor.

In 1924, when Ma married Mr. Gunness, he had 22 pupils in the Débé CMI School. On his death in 1936, Ma left Débé and moved to Coffee Street, San Fernando, and it was there that their five children grew up. These were hard years for Ma. She continued with the cane farming started by her husband, sewed clothes, and sold goods in the market, where she was known as "schoolmistress."

Ma joined Susamachar Church in 1936, and greatly admired Rev. McDonald, the long-serving minister. When he left, the church went down, and was only revived by the appointment of Rev. Ramjit. In the 1940s, Ma lived at the Papourie Road corner in Duncan Village, and when George and Myrtle married, they lived with her. Then she moved back to Coffee Street, buying the land for her house in Westwood Street, and building her home behind the Todd Street *mandir* in 1959–60. Shortly after, George and Myrtle built their own house on an adjacent plot.

Ma knows a bit about Hindu and Muslim customs, but not too much; she finds herself between Christianity, Hinduism, and Islam. She started to wear the *oronhi* about two weeks before her marriage—"You must be married!" Ma feels strongly Indian: "I like my own Indian people: my heart goes out to them quicker than anyone else." She doesn't eat pork, but since childhood has eaten ham.

Blacks look down on Indian people and call them coolie. She finds some blacks spongers, but so are Indians, too. "I have one or two very good black friends, and some of the cultured ones are very nice." "I would not like that intermarriage at all. Chinese and European are OK. But I could not stop it. I would not have too much to do with them. I am proud to be a coolie."

"I have one or two black friends, who are very good, but I keep just a little away from them. I can't be like family with them, though they may have better qualities than my Indians." She also has Chinese friends: but "those that are really close to me are my own people."

Friday, September 4

Gillian: *Preparations for the Seukeran wedding*

Preliminaries were under way on the eve of the Seukeran wedding. Mrs. Seukeran and friends had prepared food during the day—rice, curried *aloo* (potatoes) and vegetables, *dal*, and *dosti roti*.[70] There was no meat: strictly vegetarian and Hindu food—but plenty of rum. A galvanized-and-bamboo tent had been erected at the side of the house. We arrived at 8.30 p.m. to find hardly anyone there.

The singers, five women from Gasparillo, of whom, I think, two were Muslim, arrived in great form, armed with a drum, a *dhantal*, shakers, and an ancient *lotah* to be struck with a coin. They settled themselves down on a colored sheet on the floor, and began by singing five wedding songs without musical accompaniment. These songs were based on the Rama/Sita theme. Then they broke into musical accompaniment and "hotter" numbers—Hindi songs interspersed with raucous laughter.[71]

The noise was overpowering, but the atmosphere and rhythm captured everyone's attention. Little girls stood up and danced in a unique mixture of oriental dance and jive. They were followed by older women and some of the singers, all of whom wiggled their bottoms in a most uninhibited and un-Trinidad-Indian way. Why do Indians pour such scorn on black "fêters," and dancing, only to do the same themselves, but in a closer, more confined context?

A black neighbor was present. It was evident that she visits the house only on very special occasions, and had not been in for a year or more. She took no *roti*, and ate with a spoon. Everyone else used their fingers. We were offered rum and coke to drink. It was strangely incongruous to be offered home-brewed coffee in dainty little cups after tucking into a *roti* meal with our fingers.

The best man is the groom's cousin, who has flown back specially from London. He is the son of the pundit at Palmyra (the one who went into the Hanuman trance). This pundit is married to Carl Seukeran's sister. "Lionel the legislator" is flying in from London today for the wedding. The bride made no appearance at all, and the groom came in at about 11 p.m. Carl has told us again and again how his family has been steeped in Hindu "culture" for the last 5000 years. Carl "just can't live without it. It's a whole way of life."

The women's curse songs were something we had not witnessed before. It is traditional for the women to curse the father of the *doolaha* (groom) and his mother for bringing this boy into the world to steal the lovely *doolahin*. Carl was explaining the content of the songs to the groom, Merlyn, who understood nothing, and what's more, did not seem remotely interested.

Sunday, September 6

Gillian: *The Seukeran wedding*

The actual wedding was held at the Catholic Church of our Lady of Perpetual Help on Harris Promenade. The racial mixture among the congregation was immediately evident. Many of Merlyn's young Hindu and Muslim friends were there; for example, Fyzal Hydal, escorting a black girl, Mrs. Rampersad's sons, and Mahendranath Maharaj. Several women were wearing saris in the church, notably Lionel Frank's wife, Ruth, and Carl's wife. The latter wore a sari selected and bought by Carl. One bridesmaid out of the five was black. The service was inaudible, but nobody seemed to mind.

The reception was held at Oxford Club, in a bare, unpainted, undecorated room—rather like a large pavilion. The guests were welcomed by Carl and his wife, but not by the bride and groom, who appeared later. Each guest was handed a glass of champagne on arrival, though it was never made clear that it was for the toasts. We sat around awkwardly for a seemingly long time while the guests assembled and a tape recorder was found and fixed up. Mrs. Ruth Seukeran, a former San Fernando town councilor, her daughter Ria, and her husband, Peter May, and several Seukeran grandchildren sat on one side. "Lionel the legislator" sat among the mass of the people, away from his family.

Radhica Saith (Lionel's daughter) told us that she and Lennie were married according to three different religious ceremonies: Presbyterian, Sanathanist, and Arya Samaj. Her husband, Lennie Saith,[72] is Arya Samaj. His parents, who come from Chaguanas, were some of the first Samajists in Trinidad. Lennie is a PNM supporter. He had an Indian government scholarship to study engineering at a university to the north of Delhi, and then a Trinidad government scholarship to the University of Newcastle, United Kingdom to study road traffic engineering. He hopes to return there in a year's time to do a PhD. His present post is government engineer in the south of Trinidad. It is said that Lionel Seukeran tried his hardest to get into the PNM between 1956 and 1961, while he was the Independent Representative for Naparima.

Radhica was all set to study law in England, (like Ria, who is a barrister), when she was married to Lennie. Lionel and Lennie appear to get on well. (There are many small indications that Lionel likes to keep himself well in with the PNM.) Lionel Seukeran was stricter with Radhica, because he saw what education abroad had done for Ria. He was reluctant to take the chance with another daughter. By way of compensation, Lionel bought a gas station for Radhica, in a strategic position as you leave San Fernando for Port of Spain. Later in the conversation, we were told that Lionel married Radhica off before she could marry out of race (as her cousin now has done).

Our conversation was interrupted by the arrival of Ria. Lionel Frank is, according to his daughters "very Indian at home." "All o' we is one" during the day, but he is glad to be as Indian as possible at night. Considerable adoration was expressed for daddy, who, as usual, looked immaculate in an expensive English tweed suit. What a paradox! Ruth so eastern in her sari, and Lionel, so colonial in tweeds!

Lennie Saith has a government house. Ria, Radhica, and their parents live within one block of each other. Ian, the magistrate brother, is in Port of Spain. The whole family was brought up very much in the Indian tradition, but the girls were always encouraged to voice their opinions, and there seems to be a love of family discussion.

Within minutes of being introduced, Ria told us that she loves daddy (Lionel) and money. Money seems to be no object, and they are constantly traveling to and from the United Kingdom. As students, Ria and Ian spent more time in Scandinavia and on the continent than in England. Consequently, she knows little about England outside London. When Radhica and Lenny were engaged, she and Lenny were allowed to go to the drive-in cinema, with Ria as chaperone. Enough was said to suggest that democratic socialism doesn't go down well with Lionel; and there is every indication that he won't go up on a DLP ticket next time.

Monday to Thursday, September 7–10

Colin: *Farewells*

At the beginning of our last week in San Fernando, we were invited by Hansar and Bram to a farewell party held in our honor in the Todd Street *mandir*. It was a moving event, with more than 50 participants from the Hindu community. Food and drink were served, and we were presented by Hansar with a *rehal* that he had carved from a single piece of wood, decorated, and inscribed. How did he find the time?

During these final days of fieldwork, Bram told us that when the Prince Albert Street *mandir* was about to be broken down, the *mahant*, Ramsamooj Misir, went to law and eventually won $1,500 BWI compensation, which was paid to him by Amrod Singh and L. Jaggernauth. Eventually, Ramsamooj Misir moved in and occupied the temple. Ramsamooj Misir failed in his attempt to have himself recognized as a Brahmin during a *panchayat* held in the Venus Theatre outside San Fernando, and thereafter gave up public rituals. Bisram Gopie, Mr. Dialdas, and Binie Maharaj went to Usine, and acquired the land for the Todd Street *mandir*. The Bombay Indians go to the temple during the week to do *puja*, though a few, as we have seen, also turn up during the sandyha puja on Sunday mornings.

Bram added: "Hansar Ramsamooj would have left the temple without me 18 months ago, because of Brahmin pressure." The Women's

Service Group or Stri Sevak Sabha has the following officers: Jean Maharaj, president; Sita Maharaj, secretary; Mrs. Moon Persad, treasurer. The light over Lord Krishna in the *mandir* was suggested by Mrs. Deabi Persad. Bram and Hansar agree that both Mr. Binie Maharaj and Mr. Gopaul are opposed to Bhadase's reelection to the leadership of the Sanathan Dharma Maha Sabha.[73]

Friday, September 11

Gillian: *With George and Myrtle Sammy, St. Augustine*

We spent our last night in Trinidad with George and Myrtle Sammy on the campus of the University of the West Indies at St. Augustine, near Tunapuna, where George is a lecturer in the Chemical Engineering Department. We told them about the conversation we had had with Myrtle Ma, and her observation that she has a few black friends, but would never regard them as family. Myrtle smiled and enlarged on this. Whenever Ma is in any trouble, the person she first turns to is Mr. James, one of the few blacks in Débé (where she lived as the bride of the headmaster).

George, on the eve of our departure, revealed a hitherto unsuspected conservatism; he remarked that whenever men and women are in the same room, he feels that each sex should get together and discuss their own business. Although he had never mentioned this to Myrtle before, she didn't demur. Furthermore, George regards it as a woman's duty to look after her husband's every need, including cooling his tea for him. Among Indians, this involves pouring the tea from one cup to another without spilling it, until the liquid is sufficiently cool to drink.

Saturday, September 12

Colin: *Departure*

Hansar and Bram came all the way from San Fernando to take us the short distance from St. Augustine to Piarco Airport for our flight. As we were checking in, we encountered Bhadase Maraj, the former leader of the opposition, who was also leaving the island, though on another flight. Our companions told us that on a previous trip to the United States, rumor had it that Bhadase had been transfused with monkey blood.[74] Hansar gave us an unexpected parting gift—a bottle of Trinidad rum.

Notes

1. For further information on Hindu *sanskar* (*sanskara* rituals), see Steven Vertovec, *Hindu Trinidad: Religion, Ethnicity and Socio-Economic Change*, 1992, 202–10.
2. The hair cutting or shaving often took place during the Siparu Mai festival in Siparia.
3. Walter Annamunthodo, born British Guiana, 1921. Printer and trade unionist, assistant manager PNM Publishing Company, 1958–61 (Carlton Comma (ed.), *Who's Who in Trinidad and Tobago 1966*, 1966, 41).
4. The West Indian National Party was left-leaning (possibly Marxist), and economically and racially utopian (Selwyn Ryan, *Race and Nationalism in Trinidad and Tobago: A Study of Decolonization in a Multiracial Society*, 1972, 71–3).
5. Roy Adolphus Joseph, CBE (UK), born 1908; educated San Fernando Roman Catholic School and Naparima College, San Fernando; proprietor-merchant. Councilor San Fernando Borough Council, 1938, 1942–45; elected Member of the Legislative Council, 1941; elected Mayor of San Fernando 1946–49, elected Member of the Legislative Council 1951. Minister of Education and Social Services, 1951–56 (Carlton Comma (ed.), *op. cit.*, 1966, 140).
6. Details of the DLP convention in San Fernando on July 12–13, 1964 are given in Yogendra Malik, *East Indians in Trinidad: A Study in Minority Politics*, 1971, 145–7.
7. Alloy Lequay, DLP secretary in 1964. Subsequently, party activist and DLP party leader, 1972. National sports administrator of table tennis and cricket, retired 2005.
8. Hari Persad Singh, born 1905; educated Naparima College, San Fernando and Hoosier Institute, Indiana, USA. Merchant and journalist; editor of *The Observer* (Carlton Comma (ed.), *op. cit.*, 1966, 224).
9. Chanker Maharaj, born San Juan 1916. Proprietor of estates, island's champion wrestler; General Trustee, Sanathan Dharma Association Inc. ("Who's Who," in Murli J. Kirpalani *et al.* (eds.), *Indian Centenary Review: One Hundred Years of Progress, 1845–1945, Trinidad, BWI*, 1945, 149).
10. Capildeo's dismissal of the DLP Executive is discussed in Yogendra Malik, *op. cit.*, 1971, 143–4.
11. Kurmi is the market-gardener caste and in the Vaishya varna.
12. Indians in Trinidad follow the North Indian patterns of caste (or approximately equal caste) endogamy, and village exogamy.
13. Errol Mahabir, born 1931. Educated Tranquillity Boys' School and Queen's Royal College, Port of Spain; senior Staff Texaco Trinidad Inc. Elected San Fernando Borough Council 1959–; Mayor of San Fernando 1963–; third vice chairman Peoples National Movement (Carlton Comma (ed.), *op. cit.*, 1966, 163).

14. Rev. James Seunarine, born San Fernando, 1921. Educated San Fernando Presbyterian School, Naparima College, San Fernando, University of Toronto and Princeton Theological Seminary. Pastor Couva and Guaico, 1950–58; Principal, St. Andrew's Theological College, San Fernando 1959–60; Principal Naparima College, 1961– (Carlton Comma (ed.), *op. cit.*, 1966, 221).
15. Pelham Sloane-Seale, born 1921, educated Naparima College, San Fernando and Gray's Inn, London; barrister 1957. Legal Officer, Industrial Development Corporation (Carlton Comma (ed.), *op. cit.*, 1966, 225).
16. Although Brahmins were overrepresented in San Fernando, they accounted for less than a quarter of the Hindu sample (Colin Clarke, *East Indians in a West Indian Town: San Fernando, Trinidad, 1930–70*, 1986, 90–1).
17. Tagore College, San Fernando, located at Cross Crossing, was functioning at the time of Colin's visit in 1968.
18. Pundit Dharam Maharaj, born Picton estate, 1905; peasant proprietor and trade union leader ("Who's Who," in Murli J. Kirpalani *et al.* (eds.), *op. cit.,* 1945, 149).
19. This is borne out in Colin Clarke, *op. cit.*, 1986, 127.
20. Twenty-three Hindu castes were recorded among Christian East Indians, but only half knew their caste affiliation. In contrast to San Fernando's Hindus, Christian East Indians experienced no systematic relationship between caste and class (Colin Clarke, *op. cit.*, 1986, 96–7).
21. In San Fernando, over 40 percent of Muslims were affiliated to the Trinidad Muslim League and a similar percent to ASJA; 6 percent were members of the Kadiana reformist organization, and 9 percent did not know their sect (Colin Clarke, *op. cit.*, 1986, 98–9).
22. Anna Mahase (Jr.), daughter of Kenneth and Anna Mahase. Educated St. Augustine High School and Mt. Allison University, Canada, BSc., BEd. Principal St. Augustine High School 1962– (Carlton Comma (ed.), *op. cit.*, 1966, 165). LLD. University of the West Indies; LLD. Mt. Alison, Canada. Chaconia Medal, Gold, 1990.
23. Carl's father was Pundit Seukeran Sharma of Tableland. Carl's brother, Lionel Frank Seukeran, once claimed, "I am a Brahmin and the son of a Brahmin. Seven generations of Brahmin blood flows in my veins. I know more Sanskrit and Hindi than all the Pandits in Debe [*sic*]," cited in Yogendra Malik, *op. cit.*, 1971, 33–4.
24. Approximately one-third of Christian Indians sampled in San Fernando had parents who were Hindu (Colin Clarke, *op. cit.*, 1986, 99).
25. Our visitors may have become aware of Arya Samaj in the early 1940s, when they were in their early twenties, but missionaries visited Trinidad in 1910, 1928 and 1933, established the Arya Samaj Association in 1934, and provided the critique of orthodox

Sanathanism that led to the formation of the Sanathan Dharma Board of Control and the Sanathan Dharma Association in 1932 (Steven Vertovec, *op. cit.*, 1992, 117–20).

26. In San Fernando, Sanathanists accounted for three-quarters of the Hindus in the sample, followed by equal numbers of Arya Samajis, Seunarinis, and Kabir Panthis. Twelve percent did not know their sect. In Débé 65 percent of Hindus were Sanathanists, and the remainder were split almost equally among, Arya Samajis, Seunarinis, Kabir Panthis, and Agor Panthis (Colin Clarke, *op. cit.*, 1986, 98).
27. A veda is sacred knowledge or a book; here it refers to the four sacred books of the Hindus.
28. Stephen Maharaj, deputy leader of the Butler Party in 1950s; Butler Party member of the Legislative Council, Ortoire-Mayaro, 1950–56 and Ortoire-Moruga, 1956–60. DLP member of the House of Representatives, Princes Town, 1961–; Leader of the DLP opposition 1963–; pharmacist, Princes Town (Carlton Comma (ed.), *op. cit.*, 1966, 165). In the mid-1960s, he split from the DLP and formed the short-lived Workers and Farmers Party with Adrian Cola Rienzi, Makhan Dubé, and C. L. R. James.
29. Sir Henry Pierre, born 1904. Specialist surgeon and president of Trinidad and Tobago Red Cross Society (Carlton Comma (ed.), *op. cit.*, 1966, 191–2). Chaconia Medal, Gold, 1974.
30. Rudranath Capildeo, born Chaguanas, 1920. Educated Queen's Royal College, Port of Spain, winner Island Scholarship 1938; University College, London, Maths BSc. 1943 ("Who's Who," Murli J. Kirpalani *et al.* (eds.), *op. cit.*, 1945, 137); Maths MSc. and PhD., London University. Leader of the DLP, 1960–69; member of the House of Representatives for St. Augustine, 1961–7, and of the opposition in Parliament, 1961–63 (Carlton Comma (ed.), *op. cit.*, 1966, 67). Trinity Cross, 1969. Died, London, 1970. Stephen Maharaj took over as leader of the opposition in late 1963, so that Capildeo could commute to his lecturing post in mathematics at University College, London University, while remaining leader of the DLP.
31. Capildeo's reference to five seats requires explanation. In 1956 the PNM had won the majority of elected seats (13 out of 24), but the Legislative Council consisted of 31 members—24 elected, 5 nominated, and 2 appointed officials. To have a working majority, the PNM needed to control at least 16 seats. The Colonial Office agreed to allow the PNM to select two of the five nominated members; the other three positions would not be offered to people hostile to the PNM. Moreover, the votes of the two appointed officials would normally be available to the government—giving the PNM 20 votes (Selwyn Ryan, *op. cit.*, 1972, 166).
32. Sir Learie Constantine, born 1904. MBE (1954), knighted, 1962. Former West Indian test cricketer; solicitor's clerk, and barrister, 1954. Chairman, People's National Party 1956–61; elected PNM

member of the Legislative Council, Tunapuna, 1956–61, and served in the cabinet as minister of works and transport. From 1962 to 1964, High Commissioner for Trinidad and Tobago in the United Kingdom (Carlton Comma (ed.), *op. cit.*, 1966, 78). UK Life Peer, 1969. Trinity Cross, 1971 (posthumous). Died 1971.

33. The state of Rudranath Capildeo's health was poor in the late 1950s and the 1960s; he died in 1970, aged 50.
34. "Capildeo was no politician," observed Yogendra Malik (*op. cit.*, 1971, 107). For Malik's evaluation of the 1961 elections, see pages 109–27.
35. Ashford Sinanan, son of A. R. Sinanan, businessman of San Fernando. Presbyterian. Solicitor in San Fernando. Member of the Legislative Council, Victoria South, 1950–55; PDP/DLP member of the Legislative Council for Pointe-à-Pierre, 1956–61; DLP member of the House of Representatives, Siparia, 1961–66 (Carlton Comma (ed.), *op. cit.*, 1966, 224). Sinanan was a recipient of the Chaconia Medal, Gold, 1990.
36. Capildeo had been a lecturer in mathematics at the University of the Sudan.
37. Vishnudat Datta Maharaj, born 1905, Diamond Village. Provision merchant. Pundit and president, Arya Pratinidhi Sabha ("Who's Who," in Murli J. Kirpalani *et al.* (eds.), *op. cit.*, 1945, 149).
38. Ada Date-Camps, born Grenada. San Fernando medical practitioner and PNM senator, 1962– (Carlton Comma (ed.), *op. cit.*, 83). Dr. Date-Camps remained a PNM senator until 1971, and was vice president of the Senate in 1970–71. Awarded the Chaconia Medal, Gold, 1981.
39. Murli J. Kirpalani *et al.* (eds.), *op. cit.*, 1945.
40. Similar to Chamar, one of the lowest Sudra castes in Trinidad.
41. Mohammed Yakub Khan, born 1892, store manager; vice president, Tackveeyatul Islamic Association ("Who's Who," in Murli J. Kirpalani *et al.* (eds.), *op. cit.*, 1945, 143).
42. Syed Abdul Aziz came to Trinidad from Afghanistan as an indentured laborer in 1883, and settled in Iere Village, Princes Town. Appointed the first *kazi* or Muslim chief judge in 1907; with Haji Ruknuddin he founded the Tackveeyatul Islamic Association in 1926, but died a year later. He was also one of the founders of the East Indian National Association of Princes Town (Kris Rampersad, *Finding a Place: IndoTrinidadian Literature*, 2002, 56).
43. Abdool Gany, born 1889 at Maracas; merchant; president, Tackveeyatul Islamic Association and Sunnat-ul-Jamaat Association ("Who's Who," in Murli J. Kirpalani *et al.* (eds.), *op. cit.*, 1945, 139).
44. Hon. George Fitzpatrick (1875–1920), first East Indian nominated to the Legislative Council in 1912.
45. Dr. Frank Mahabir, born Princes Town, 1888; first Trinidad Indian to qualify as a physician and surgeon and join the civil service;

district medical officer in various localities ("Who's Who," in Murli J. Kirpalani *et al.* (eds.), *op. cit.*, 149).

46. Jules Mahabir, born Princes Town 1891. Educated Naparima College, San Fernando and Queen's Royal College, Port of Spain, Gray's Inn, London; barrister (1917) and magistrate ("Who's Who," in Murli J. Kirpalani *et al.* (eds.), *op. cit.*, 1945, 147).
47. Adrian Cola Rienzi (aka Krishna Deonarine until 1927); born Palmyra, 1905. Educated Naparima College, San Fernando, Trinity College Dublin, and the Middle Temple, London; called to the Bar, 1937. Founder and leader of the oil, sugar, and transport-workers unions in the late 1930s; mayor of San Fernando 1939–42; elected member of the Legislative Council for County Victoria, 1937–44; member of the Governor's Executive Council, 1941–. He gave up his political and union careers to work as a Crown Counsel in 1944 ("Who's Who," in Murli J. Kirpalani *et al.* (eds.), *op. cit.*, 1945, 161). By 1964 Rienzi was assistant solicitor general and lived in St. Clair, the elite neighborhood of Port of Spain (Carlton Comma (ed.), *op. cit.*, 1966, 205–6). He was a cofounder with Makhan Dubé, C. L. R. James, and Stephen Maharaj of the multiracial Workers and Farmers Party, which was routed in the 1966 elections. He died in 1972.
48. God of Truth.
49. Sun God.
50. More than 50 percent of Hindu survey respondents in San Fernando and 60 percent in Débé lived in houses with *jhandi* (Colin Clarke, *op. cit.*, 1986, 101).
51. Mohammed Shah, born California 1908. Law clerk, court interpreter, and auctioneer. Prominent in scout movement ("Who's Who," in Murli J. Kirpalani *et al.* (eds.), *op. cit.*, 1945, 165). He defeated "Buzz" Butler in the 1958 Federal Election in St. Patrick (Yogendra Malik, *op. cit.*, 1971, 101).
52. In 1961 the DLP won 10 seats out of 30 contested.
53. Trinidad term for corruption or fraud, see Lise Winer, *Dictionary of the English/Creole of Trinidad and Tobago*, 2009.
54. "Jack" (John Bryan) Kelshall, born 1912, educated St. Mary's College; solicitor in San Fernando (Carlton Comma (ed.), *op. cit.*, 1966, 141). A white Creole, left-wing politician, he operated behind the scenes in Trinidad politics from the 1940s, but without electoral success.
55. For an account of the birth-control issue in the 1956 election, see Selwyn Ryan, *op. cit.*, 1972, 151–5.
56. Medical doctor. PNM president of the Trinidad and Tobago Senate, 1971–87. Trinity Cross, 1977.
57. These family data suggest that 43 percent of the San Fernando elite were Creole, and 57 percent East Indian in 1964. This East Indian proportion is closer to Frank Cleghorn's version (that 66 percent of the elite was East Indian) than the figure derived from Carlton

Comma's *Who's Who in Trinidad and Tobago, 1966*, which gave the Creole proportion as 72 percent. It seems clear that while the majority of the old elite was Creole, East Indians were closing fast, and that the town's elite at independence was racially and culturally split in half.

58. (Albert) Gerard Montano, born 1918. Educated Naparima College and St. Benedict's College, San Fernando. Executive, managing director Imperial Stores, San Fernando. Former councilor and mayor of San Fernando; elected member of the Legislative Council, San Fernando East 1956–61, and member of the House of Representatives, San Fernando East 1961–70; minister of housing and local government 1956–61; minister of works 1961–64; minister of home affairs and local government 1964–70 (Carlton Comma (ed.), *op. cit.*, 1966, 177). Ambassador Extraordinary and Plenipotentiary, Trinidad and Tobago, 1970–. Trinity Cross, 1979. Died 2005.
59. Frank Ramnarine Misir, educated Princes Town Presbyterian School and Naparima College, San Fernando; called to the Bar, Gray's Inn, London, 1948. Employed at U.S. base Chaguanas, 1941–43 (Carlton Comma (ed.), *op. cit.*, 1966, 173). Frank Misir had a distinguished career as a barrister in Trinidad and Tobago, becoming a Queen's Counsel in the late 1960s, and judge of the high court in 1984.
60. Ajodha Singh, born Roussillac, south of San Fernando, 1906; attended Naparima College, San Fernando. A law clerk based in San Fernando and the owner of cocoa estates, he also provided cures as a masseur free of charge ("Who's Who," in Murli J. Kirpalani *et al.* (eds.), *op. cit.*, 1945, 167). Largely as a result of his fame as a healer, he was elected member of the Legislative Council in 1950 and reelected in 1956. Died 1961.
61. Lord David Pitt, born Grenada, 1913. Medical doctor, 1938; practiced in Trinidad. Founder (1943) West India National Party; member of the San Fernando Borough Council and deputy mayor. Went to the United Kingdom in 1947, and had a successful career as a general practitioner in London, and as an elected member of the London County Council and the Greater London Council. Appointed a Life Peer in 1975, he was awarded the Trinity Cross, 1976. Died 1994.
62. Dr. Patrick Solomon, born 1910. Educated Tranquillity Boys, St. Mary's College, Port of Spain, University College, London, and Queen's University, Belfast. Medical practitioner and politician. Elected member of the Legislative Council, Port of Spain, South 1946–56, deputy leader PNM 1956–; Port of Spain, West, 1956–; minister of education and culture, 1956–59, minister of home affairs, 1959–64, deputy prime minister and minister of external affairs, 1964–66 (Carlton Comma, ed., *op. cit.*, 1966, 27). Trinidad and Tobago representative to the United Nations, 1966–71, and London, 1971–76. Trinity Cross, 1978. (Patrick Solomon, *Solomon: An Autobiography*, Port of Spain, Inprint, 1981). Died 1997.

63. McDonald Moses, born St. Lucia, 1910. Educated London University. Cofounder and vice president Oilfield Workers Trade Union, 1937–48; cofounder and president All Trinidad Sugar Estates and Factories Workers Trade Union, 1942–48; cofounder and trustee, Trinidad and Tobago Trades Unions Congress, 1938–44; member of Commission on Subversive Activities, 1963–64 (Carlton Comma (ed.), *op. cit*, 1966, 177).
64. Ranjit Kumar, born 1912, BSc. and chartered civil engineer. Assistant city engineer, Port of Spain. Came to Trinidad from India in 1935 as distributor of first Indian film, *Bala Joban* ("Who's Who," in Murli J. Kirpalani *et al.* (eds.), *op. cit.*, 1945, 145). Chairman of the Victoria County Council, 1947–50; elected member of the Legislative Council, 1946–56 (Carlton Comma (ed.), *op. cit.*, 1966, 146).
65. County Victoria is the administrative division surrounding and including San Fernando.
66. Party formed by Tubal Uriah "Buzz" Butler. Butler was born in Grenada, 1897. Employed in Trinidad oilfields 1921–30; leader of oilfield strikes 1935–37; interned for security purposes 1939–45. Elected member of Legislative Council for St. Patrick West in 1950–61. Defeated in federal election of 1958 by Mohammed Shah (Carlton Comma (ed.), *op. cit.*, 1966, 63–4). Trinity Cross, 1970. Died 1977.
67. Mitra Sinanan, son of A. R. Sinanan, San Fernando businessman, born 1910. Presbyterian, educated Naparima College, San Fernando, the London School of Economics and Middle Temple, London—barrister and Queen's Counsel. Defended Butler in court in 1937 after the oilfield strikes. Member of Legislative Council, Caroni South, 1950–56; acted as deputy speaker, leader of the opposition, minister of education, minister of works, and minister of health 1950–55; PDP/DLP member of the Legislative Council, Caroni Central, 1956–61 (Carlton Comma (ed.), *op. cit.*, 1966, 63–4). Trinity Cross, 1979.
68. Latchan Jaggernauth, born 1899, San Fernando. Sportsman, cricket enthusiast, and president of Rahamut Cricket Cup Competition ("Who's Who," in Murli J. Kirpalani *et al.* (eds.), *op. cit.*, 1945, 141). OBE (UK).
69. French for aunt, see Lise Winer, *op. cit.*, 2009.
70. *Dosti roti* consists of two layers, which are rolled out and stuck together with butter before being cooked. See Lise Winer, *op. cit.*, 2009.
71. This is chutney music, brought to Trinidad by Indian indentured laborers, and performed by women for women before Hindu weddings. The themes focus on sex, rudeness, and vulgarity. Chutney is now performed in public, and is influenced by soca. Since the 1980s, men and women have been singing chutney for prizes, mixing Hindi

and English lyrics, and engaging in large-scale dancing events—not unlike jumping-up to calypsos at Carnival. See Selwyn Ryan, *The Jhandi and the Cross: The Clash of Cultures in Post-Creole Trinidad and Tobago*, 1999, 176–7.

72. Dr. Lennie Saith is a PNM senator, and held a series of ministerial portfolios in the early 1990s and the 2000s, the current one as minister of energy and energy industries.
73. However, Bhadase Maraj continued as leader of the Sanathan Dharma Maha Sabha until his death in 1971.
74. Monkey blood is a far-fetched, Trinidad idea—in this case attempting to explain Bhadase's erratic behavior. In reality, Bhadase became sick and bedridden in 1959, and a recent study claims that for a while he became addicted to the painkiller pethidine (Kirk Meighoo, *Politics in a "Half-Made" Society: Trinidad and Tobago, 1925–2001*, 2003, 52).

References

Braithwaite, Lloyd, "Social Stratification in Trinidad," *Social and Economic Studies*, 2, nos. 2 and 3 (1953): 5–175.

Braithwaite, Lloyd, "Social Stratification and Social Pluralism," in Vera Rubin (ed.), "Social and Cultural Pluralism in the Caribbean," *Annals of the New York Academy of Sciences*, 83 (1960): 816–36.

Clarke, Colin, *East Indians in a West Indian Town: San Fernando, Trinidad, 1930–70,* London Research Series in Geography, 12 (London: Allen and Unwin, 1986).

Clarke, Colin, "Society and Electoral Politics in Trinidad and Tobago," in Colin Clarke (ed.) *Society and Politics in the Caribbean* (London: St. Antony's/Macmillan, 1991), 47–77.

Clarke, Colin, "Spatial Pattern and Social Interaction among Creoles and Indians in Trinidad and Tobago," in Kevin Yelvington (ed.), *Trinidad Ethnicity* (London: Warwick-Macmillan Caribbean Studies, 1993), 116–35.

Comma, Carlton, *Who's Who in Trinidad and Tobago, 1966* (Port of Spain: Caribbean Publishers, 1966).

Frank, H. A., *Roaming through the West Indies* (New York: Century Co., 1923).

French, Patrick *The World Is What It Is: The Authorized Biography of V. S. Naipaul* (London: Picador, 2008).

Ghany, Hamid, *Kamal: A Lifetime in Politics, Religion and Culture* (San Juan: Kamaluddin Mohammed, 1996).

Goodenough, Stephanie, "Race, Status and Residence, Port of Spain, Trinidad," Ph.D. thesis (Liverpool: University of Liverpool, 1976)

Hall, Douglas, *A Man Divided: Michael Garfield Smith* (Kingston, Jamaica: University of the West Indies Press, 1997).

Herskovits, Melville and Frances Herskovits, *Trinidad Village* (New York: Octagon Books, (1964 edn.).

Khan, Aisha, *Callaloo Nation: Metaphors of Race and Religious Identity among South Asians in Trinidad* (Durham and London: Duke University Press, 2004).

Kirpalani, Murli J., Mitra G. Sinanan, S. M. Rameshwar, and L.F. Seukeran (eds.), *Indian Centenary Review, One Hundred Years of Progress,*

1845–1945, Trinidad, B.W.I., Indian Centenary Review Committee (Port of Spain: Guardian Commercial Printery, 1945).
Klass, Morton, *East Indians in Trinidad: A Study of Cultural Persistence* (New York: Columbia University Press, 1961).
Klass, Morton, *Singing with Sai Baba: The Politics of Revitalization in Trinidad* (Prospect Heights, Illinois: Waveland Press, 1991).
Macedo, Lynne, *Fiction and Film: The Influence of Cinema on Writers from Jamaica and Trinidad* (Chichester: Dido, 2003), 33.
Mahabir, Winston, *In and Out of Politics* (Port of Spain: Inprint, 1978).
Malik, Yogendra K. *East Indians in Trinidad: A Study in Minority Politics* (London: Oxford University Press for Institute of Race Relations, 1971).
Meighoo, Kirk, *Politics in a "Half-Made" Society: Trinidad and Tobago, 1925–2001* (Oxford: James Curry; Kingston, Jamaica: Ian Randle Publishers; Princeton: Marcus Wiener Publishers, 2003).
Miller, Daniel, *Modernity: An Ethnographic Approach. Dualism and Mass Consumption in Trinidad* (Oxford: Berg, 1997).
Morton, Sarah E., *John Morton of Trinidad* (Toronto: Westminster, 1916).
Munasinghe, Viranjini, *Callaloo or Tossed Salad: East Indians and the Cultural Politics of Identity in Trinidad* (Ithaca and London: Cornell University Press, 2001).
Naipaul, V.S., *The Mystic Masseur* (London: André Deutsch, 1957).
Naipaul, V.S., *A House for Mr Biswas* (London: André Deutsch, 1961).
Naipaul, V.S., *The Middle Passage: Impressions of Five Societies—British, French and Dutch—in the West Indies and South America* (London: André Deutsch: 1962).
Nicholls, David, "East Indians and Black Power in Trinidad," *Race*, 12 (1971): 443–59.
Niehoff, Arthur and Juanita Niehoff, *East Indians in the West Indies*, Publications in Anthropology, no. 6 (Milwaukee: Public Museum, 1960).
Premdas, Ralph, "Ethnic Conflict in Trinidad and Tobago: Domination and Reconciliation," in Kevin Yelvington (ed.), *Trinidad Ethnicity* (London: Warwick University-Macmillan Caribbean Series, 1993), 136–60.
Ramchand, Ken, *Mr Speaker, Sir: An Autobiography of Lionel Frank Seukeran* (San Fernando: Chandrabose Pub., 2006).
Ramesar, Marianne, "The Repatriates" in David Dabydeen and Brinsley Samaroo (eds.), *Across the Dark Waters: Ethnicity and Indian Identity in the Caribbean* (London: Warwick University-Macmillan Caribbean Series, 1996), 175–200.
Rampersad, Kris, *Finding a Place: IndoTrinidadian Literature* (Kingston, Jamaica: Ian Randle Publishers, 2002).
Ramsamooj, Hansar, *A Collection of Poems* (South Oropouche, Trinidad: Self-published, 1985.)
Richardson, Bonham, "Livelihood in Rural Trinidad in 1900," *Annals of the Association of American Geographers*, 65 (1975): 240–51.
Rubin, Vera, "Culture, Politics and Race Relations," *Social and Economic Studies*, 11 (1962): 433–55.

Ryan, Selwyn, *Race and Nationalism in Trinidad and Tobago: A Study of Decolonization in a Multi-Racial Society* (Toronto: University of Toronto Press, 1972).

Ryan, Selwyn (ed.), *Trinidad and Tobago: The Independence Experience, 1962–1987* (St. Augustine, Trinidad: Institute of Social and Economic Research, University of the West Indies, 1988).

Ryan, Selwyn, *The Disillusioned Electorate: The Politics of Succession in Trinidad and Tobago* (Port of Spain, Trinidad: Inprint Caribbean Ltd., 1989).

Ryan, Selwyn (ed.), *Social and Occupational Stratification in Contemporary Trinidad and Tobago* (St. Augustine, Trinidad: Institute of Social and Economic Research, University of the West Indies, 1991).

Ryan, Selwyn, *The Jhandi and the Cross: The Clash of Cultures in Post-Creole Trinidad and Tobago* (St. Augustine, Trinidad: Sir Arthur Lewis Institute of Social and Economic Studies, University of the West Indies, 1999).

Ryan, Selwyn, *Deadlock! Ethnicity and Electoral Competition in Trinidad and Tobago, 1995–2002* (St. Augustine, Trinidad: Sir Arthur Lewis Institute of Social and Economic Studies, University of the West Indies, 2003).

Ryan, Selwyn and Taimoon Stewart (eds.), *Power: The Black Power Revolution 1970: A Retrospective* (St. Augustine, Trinidad: Institute of Social and Economic Research, University of the West Indies, 1995)

Selvon, Samuel, *A Brighter Sun* (London: Alan Wingate, 1952).

Selvon, Samuel, *Turn Again Tiger* (London: MacGibbon and Kee, 1958).

Solomon, Patrick, *Solomon: An Autobiography* (Port of Spain: Inprint, 1981).

Varadarajan, Tunku, "The Strangest Thing about Trinidad," *Oxford Magazine*, fourth week, Trinity Term (1992): 14.

Vertovec, Steven, *Hindu Trinidad: Religion, Ethnicity and Socio-Economic Change* (London: Warwick University-Macmillan Caribbean Studies, 1992).

Vertovec, Steven, *The Hindu Diaspora: Comparative Patterns* (London: Routledge, 2000).

Winer, Lisa, *Dictionary of the English/Creole of Trinidad and Tobago* (Montreal, Kingston, London and Ithaca: McGill-Queen's University Press, 2009)

Wood, Donald, *Trinidad in Transition* (London and New York: Oxford University Press, 1968).

Index